D0389857

# NOTES FROM AN
# APOCALYPSE

ALSO BY MARK O'CONNELL

*To Be a Machine*

# NOTES FROM AN APOCALYPSE

## A Personal Journey to the End of the World and Back

MARK O'CONNELL

www.doubleday.com



*Jacket illustration by María Medem*
*Jacket design by John Fontana*

Library of Congress Cataloging-in-Publication Data

Names: O'Connell, Mark, [date] author.
Title: Notes from an apocalypse : a personal journey to the end of the world and back / Mark O'Connell.
Description: First edition. | New York : Doubleday, [2020]
Identifiers: LCCN 2019024066 (print) | LCCN 2019024067 (ebook) | ISBN 9780385543002 (hardcover) | ISBN 9780385543019 (ebook)
Subjects: LCSH: Survivalism. | Emergency management—Social aspects.
Classification: LCC GF86 .O36 2020 (print) | LCC GF86 (ebook) | DDC 613.6/9—dc23
LC record available at https://lccn.loc.gov/2019024066
LC ebook record available at https://lccn.loc.gov/2019024067

MANUFACTURED IN THE UNITED STATES OF AMERICA
1 3 5 7 9 10 8 6 4 2

First Edition

*For Amy, Mike, and Josephine*

Adults keep saying: "We owe it to the young people to give them hope." But I don't want your hope. I don't want you to be hopeful. I want you to panic. I want you to feel the fear I feel every day. And then I want you to act. I want you to act as you would in a crisis. I want you to act as if our house is on fire. Because it is.

—*Greta Thunberg*

These times of ours are ordinary times, a slice of life like any other. Who can bear to hear this, or who will consider it?

—*Annie Dillard*

# CONTENTS

1. Tribulations . . . . . . . . . . . . . . . . . . . . 1

2. Preparations . . . . . . . . . . . . . . . . . . . . 19

3. Luxury Survival . . . . . . . . . . . . . . . . . 43

4. Bolt-hole . . . . . . . . . . . . . . . . . . . . 71

5. Off-World Colony . . . . . . . . . . . . . . 102

6. Under the Hide. . . . . . . . . . . . . . . . 132

7. The Final Resting Place of the Future . . . 183

8. The Redness of the Map . . . . . . . . . . 221

*Acknowledgments* . . . . . . . . . . . . . . . . 255

# NOTES FROM AN
# APOCALYPSE

# 1
# TRIBULATIONS

It was the end of the world, and I was sitting on the couch watching cartoons with my son. It was late afternoon, and he was sprawled across my lap, looking at a show about a small Russian peasant girl and the comic scrapes she gets embroiled in with her long-suffering bear companion. I was holding my phone over his head, scrolling downward through my Twitter feed. The bear and the girl were involved in some kind of fishing-based slapstick escapade, in which the bear was doing a lot of stumbling about and falling over. My son was giggling happily at this, turning his face periodically upward to ensure that I was aware of the amusing pratfalls unfolding on our television screen.

On the smaller screen of my phone, I came across an embedded YouTube video on which, precisely because its accompanying text advertised it as "soul-crushing" and "heart-wrenching," I clicked without hesitation.

As my son watched his cartoon, I held my phone above his line of vision and watched an emaciated polar bear dragging itself across a rocky terrain, falling to its knees and struggling to lift itself again, hauling its tufted carcass onward toward a cluster of rusting metal barrels half filled with trash, from which it eventually managed to paw out what looked like a knuckle of raw bone, more or less totally devoid of meat. The animal was a pathetic sight; because of the wasting effects of malnutrition, it looked more like a gargantuan stoat or weasel than a polar bear. As it slowly chewed whatever it was that it had managed to scavenge from the trash, its eyes half closed in deep and terminal fatigue, a white tide of saliva frothed slowly from its mouth, while over this footage a cello played a slow and mournful glissando.

I turned down the sound on my phone so as not to attract my son's attention, his inexorable questions. He was three then, and our relationship in those days took the form of an endless interrogation.

A text at the bottom of the screen explained that the footage was shot near an abandoned Inuit village in the northern Canadian tundra, where the bear had strayed in search of food, the population of seals, its usual food source, having been drastically diminished by the effects of climate change.

My soul remained uncrushed, my heart more or less unwrenched. I felt instead a creeping disgust at the footage itself, at the manner of its presentation—the lachrymose music, the stately pace of the editing—which seemed designed to elicit in me a recognition of my own contribution to this terrible situation, together with a virtuous and perhaps even redemptive swelling of sorrow, of noble sadness at the ecological destruction in which I myself was playing a role. It occurred to me then that the disgust I felt was the symptom of a kind of moral vertigo, resulting from the fact that the very technology that allowed me to witness the final pathetic tribulations of this emaciated beast was in fact a cause of the animal's suffering in the first place. The various rare-earth minerals that were mined for the phone's components in places whose names I would never be required to learn; the fuels consumed in the course of its construction, its shipping halfway across the world, its charging with electrical current on a daily basis: it was for the sake of all this, and in my name, that the bear was starving and dragging itself across the rocky ground.

The slapstick capering of the cartoon bear my son was watching on the television screen and, above his head, the awful distress of the real bear on the smaller screen: the absurd juxtaposition of these images, simultaneously summoned from the ether and vying for attention, generated a strange emotional charge, a surge of shame and sadness at the world my son would be forced to live in, a shame and sadness that I in turn was passing on to him.

It seemed to me that I was being confronted with an impossible problem: the problem of reconciling the images on these two screens, or at least of living with the fact of their irreconcilability. The bears in his world were always hanging out with kids and having adventures, living in cabins, enduring comic mishaps, coming good in the end. The bears in mine were all rummaging in bins and starving to death. I wanted him to live in that first world, that good world, as long as possible, but I knew that soon enough he would have to leave it and live in the future. And it was not obvious to me how a person was supposed to raise children, to live and work with a sense of meaning and purpose, in the quickening shadow of that future.

It didn't take much, in those days, to set me off on a path toward the end of the world. There were frequent opportunities to indulge my tendency toward the eschatological. Cartoons, viral videos, radio news bulletins, uneasy exchanges with neighbors about how it never used to be this warm in February. So many things felt like a flashback sequence in the first act of a postapocalyptic movie, like we were living right before the events of the main timeline kicked in. I knew that this kind of thinking was as old as human civilization itself, that imagining the apocalypse was immemorially a response to times of rapid change and uncertainty. This recognition made it no less oppressive, no less real.

What did I feel when I thought about my son and his future? I felt a kind of abstract but all-consuming melancholy. My love for him felt like an insoluble moral problem. The smart money seemed to be on apocalypse, but as a parent I felt I had some kind of moral duty to be deluded about the future, to avert my gaze from the horizon. I was by no means living up to this duty.

We are alive in a time of worst-case scenarios. The world we have inherited seems exhausted, destined for an absolute and final unraveling. Look: there are fascists in the streets, and in the palaces. Look: the weather has gone uncanny, volatile, malevolent. The wealth and power of nominal democracies is increasingly concentrated in the hands of smaller and more heedless minorities, while life becomes more precarious for ever larger numbers of people. The old alliances, the postwar dispensations, are lately subject to a dire subsidence. The elaborate stage settings of global politics, the drawing rooms and chandeliers, are being dismantled, disappearing off into the wings, laying bare the crude machinery of capital. The last remaining truth is the supreme fiction of money, and we are up to our necks in a rising sludge of decomposing facts. For those who wish to read them, and for those who do not, the cryptic but insistent signs of apocalypse are all around.

Another browser window, another omen of the end. A UN report detailing how one million species are at risk of imminent extinction. An image of a waterfall cascading into the Arctic from the sheer cliff face of a melting glacier. The proliferation of antibiotic-resistant diseases. And all of it subject to the great flattening effect of the online discourse.

Listen. Attune your ear to the general discord, and you will hear the cracking of the ice caps, the rising of the waters, the sinister whisper of the near future. Is it not a terrible time to be having children, and therefore, in the end, to be alive? This question is not a rhetorical one. I myself go back and forth on it, obsessively, helplessly, talking myself in and out of different kinds of answers. And if it is now a terrible time to be alive and having children, you have to ask when, in the scheme of things, it was ever a good one.

Having children is the most natural thing in the world, and

at the same time among the most morally fraught. During the time I am talking about here, I was consumed—pointlessly, morbidly consumed—by the question of whether having brought a human being into the world was a terrible ethical blunder, given what seemed to lie ahead. The last thing the world needed, after all, was more people in it, and the last thing any hitherto nonexistent person needed was to be in the world. It was of course a little late in the day now, the deed being well and truly done, to be giving serious attention to these fundamental questions, but then again it was precisely the day's lateness that brought the questions themselves into absolute focus.

Because the first thing to be said about becoming a parent, whether it happens by choice or by chance, is that it is one of only very few events in life that are entirely irreversible. Once you're in, existentially speaking, you're in. And so the real question, the only question—given what the world is, how dark and uncertain its future—is that of how to proceed. How are we supposed to live, given the distinct possibility that our species, our civilization, might already be doomed?

Should we just ignore the end of the world?

Again, the question is not wholly ironic: on a personal level, I'm open to the suggestion that such a response—by which I mean no response at all—might well be the sanest one, given the situation. It's certainly the easiest response, and therefore by some distance the most tempting. The problem, our problem as a culture—which, I may as well admit, neatly dovetails with my problem as a writer, and to some hopefully limited extent yours as a reader—is one of boredom.

Because let's at least be honest about this: the apocalypse is profoundly dull. I for one am sick to death of the end of days. I'm sick, in particular, of climate change. Is it possible to be terrified and bored at the same time? Is it possible, I mean, to be bored of terror—if not of the kind of literal terror that privileged people like me rarely experience, at least the kind of abstract terror that is released like a soporific gas from the whole topic of ecological catastrophe?

The threat of nuclear war that hung over much of the twentieth century at least had the advantage of focusing the mind. Nuclear war, for all its considerable flaws, you at least have to admit was gripping. It adhered to certain established narrative conventions. You had near misses, global panics. You had mutually assured destruction, game theory, mushroom clouds, total and instant annihilation. You had plot, was what you had: you had drama. And even more crucially, you had characters. You had protagonists and antagonists, guys with fingers on buttons who either did or did not choose to push them. And when it came to protestors on the streets calling for complete nuclear disarmament, you had an entirely rational and achievable demand. As imminent as it surely seemed for so long, nobody actually wanted nuclear war. Everyone understood that it would have been an act of madness, an obvious moral grotesquerie, to punch in the codes, to launch the warheads, to cause annihilation of an unprecedented swiftness and scale.

And the important point in this context is that we ourselves were not among the protagonists and antagonists. We were not going to be punching in any launch codes either way. We were bystanders, whose role was limited to cowering in terror, maybe holding the occasional placard, partaking here and there in a chant if called upon to do so. We didn't want to go to our graves, but at least we knew that if we did, we would do so more or less passively, more or less without blame.

"Once people saw in the apocalypse the unknowable avenging hand of God," as the German writer Hans Magnus Enzensberger put it in an essay from the late 1970s. "Today it appears as the methodically calculated product of our own actions." The apocalypse, he writes, "was also once a singular event, to be expected unannounced as a bolt from the blue: an unthinkable moment that only seers and prophets could anticipate—and, of course, no one wanted to listen to their warnings and predictions. Our end of the world, on the other hand, is sung from the rooftops even by the sparrows; the element of surprise is missing; it seems only to be a question of time. The doom we picture for ourselves is insidious and torturingly slow in its approach, the apocalypse in slow motion."

This slowness is something new in history. The apocalypse, in both its religious and its secular modes, has always appeared in the form of a blinding flash, a sudden intercession of divine or technological power. There is no mythological template to help us make sense of the current mutated form of the end times. We don't know how to think about it, how to give it the form of myth and story, and so it metastasizes and spreads, a blood sickness in the culture. The slow and insidious doom identified by Enzensberger takes multiple forms, insinuating itself into the most unlikely of places. There is no one cause, no single locus of apocalyptic unease. It's all horsemen, all the time.

And there is a deeper register to this truth: I don't know that I would have it any other way. I want my toilet to flush. I want streaming music. I want to buy what I want to buy, eat what I want to eat, go where I want to go. I want to be able to leave my tiny island in the North Atlantic when I need to, or just when I feel like it. And if polar bears are going to be starving to death due to habitat destruction, I want to be able to watch deeply upsetting YouTube videos about it.

You will note that this book about the apocalyptic tenor of our time features a great many interludes of travel to distant places—to Ukraine and California and South Dakota, to the highlands of Scotland and to New Zealand—and that I neither walked nor sailed nor took a train to any of these places. And let the record further show that during the time I was traveling for this book, I was also traveling to many other places to talk about my previous book. My footprint is as broad and deep and indelible as my guilt.

My days are a procession of last things, seals opened. I myself am the apocalypse of which I speak. That is the prophesy of this book.

The origins of my own obsession with this topic are buried deep in the submerged civilization of childhood. One of my earliest memories pertains to the end of the world. I was in the kitchen of my grandmother's house. My grandmother was at the stove, and I was sitting at the table with my uncle, who was explaining to me the likely outcome of a nuclear exchange between the United States and Russia. I don't know why he was doing this, other than that it was the mid-eighties and the chill of the Cold War could still be felt in the air, and he was in any case very much that kind of uncle. I'm guessing I was five, maybe six.

From a fruit bowl in the center of the table he selected first three apples, and then two small clementines. The apples he spread out on the table at equal intervals; then he carefully positioned the clementines on top of the rightward and leftward apples, leaving the middle one untouched. The apple on the left, he said, was the United States, while the one on the right was Russia. And the apple in the middle, he said, was Ireland, where we were. Its location, he explained, was more or less exactly halfway between the United States and the Soviet Union, gigantic countries of which there were two important things to be said: that they despised each other for reasons too complicated to get into with a child, and that they were both armed to the teeth with nuclear missiles, bombs so powerful that they were capable of wiping out entire countries in a matter of moments. I should imagine, he said, that the clementines were nuclear bombs.

Let's say the Russians decided to launch a missile at the Americans, he said. The missile's launch would be quickly picked up by the Americans' detection systems, and they would immediately launch their own missile in retaliation. Here he plucked the clementines from the top of the superpower apples and sent them arcing through the air above the table. And because Ireland was located exactly between Russia and the United States, he said, the missiles would collide right over our heads.

I remember what he said then, as he smashed the two clementines together above the middle apple, and the grim satisfaction with which he said it: "Good night, Irene!"

I don't remember how I felt about this performance. The fact that I'm telling you about it now, more than thirty years later, in the context of a book about the apocalypse, suggests that its effects registered in the psychic depths. Oddly enough, it never stopped me enjoying clementines, or "easy peelers," as they've since come to be known.

Writing in the fifth century AD, Saint Augustine observed that three centuries before his own time, the earliest followers of Jesus, consumed with apocalyptic fervor, believed themselves to be living in the "last days" of creation.

"And if there were 'last days' then," he wrote, "how much more so now!"

The point being, for our purposes here, that it has always been the end of the world. Our entire civilization—from Ragnarok to Revelation to *The Road*—rests on a foundation of flood and fire. But what if now it's *especially* the end of the world, by which I mean *even more* the end of the world: really and truly and at long last the end (or something like it)? And this in turn raises the question of what is meant by the end of the world. Because, in truth, isn't the idea an absurd one? How could the world just end? The world is not a business to be wound up, a property to be foreclosed on overnight.

Global nuclear war, it is true, could in theory wipe out all organic life on the planet, but that seems, at time of writing at least, a long shot.

As for climate change, only on the outermost edge of the spectrum of possibilities can be glimpsed the perfect black of annihilation. No, what is actually meant by the end of the world is, in its particulars, a province of terrors fleetingly glimpsed, barely apprehended. What we are talking about is the collapse of the systems by which the known world operates, slowly and then all at once.

It is customary now to speak of the *looming* effects of climate change, the *looming* catastrophe. We live in a time of looming, of things impending and imminent. The culture is presided over by an unsettling array of looming phenomena—looming climate catastrophe, certainly, but also looming right-wing populism, and the looming specter of the employment crisis that will be brought on by widespread automation across multiple economic sectors.

Here's a thing you can do if you are interested in receiving portents and symbols, signatures of our time. You can go to Google's Ngram feature—a graphing application that measures, over a set period, the occurrences of a particular word or phrase in the thirty million or so volumes that have so far been scanned by Google in its effort to digitize the world's books—and you can do a search for the phrase *looming crisis,* and you will see a blue line measuring usages of the term between 1800 and 2008, and you will note that, after some barely perceptible bumps in the years around the First and Second World Wars, it begins to rise non-trivially in the early years of the Cold War, before climbing so precipitously throughout the '80s, '90s, and 2000s, with such dizzying speed and persistence, that it becomes a Matterhorn of cultural anxieties. The graph itself takes on the appearance of a looming crisis, a line of rising terror.

And yet whatever it is we might mean by the end, are we not in this sense already at the beginning of it? Has the looming not in fact given way to the crisis itself?

The above, it strikes me, is as much personal reflection as cultural observation. This sense of looming crisis was one that I felt intensely throughout the time I am writing about here. I am talking, let me tell you, about a long run of very bad days: I couldn't sneeze without thinking it was a portent of end times. I was obsessed with the future, an obsession that manifested as an inability to conceive of there being any kind of future at all. Personal, professional, and political anxieties had coalesced into a consuming apprehension of imminent catastrophe. I suppose it could be said that I was depressed—and in fact it was said, often enough, by me—but it was a state characterized not by a closure against the world, but by an excessive openness to it. There was a feedback loop in operation, whereby I perceived in the chaos of the world at large a reflection of my own subjective states, and the perception of the one seemed to heighten the experience of the other. Everything that mattered seemed poised on the brink of total collapse: my mind, my life, the world.

Another, blunter, way of putting all this would be to say that my journalistic objectivity, a fragile edifice to begin with, was under considerable strain.

I remember, during this time, awakening in the abject dawn from a nightmare: of imperious thudding on our front door, of pale hands plunging through the letter box, palpating the air we breathed, my little family and I, in our little house. These grasping fingers were not the most upsetting part of the dream. The most upsetting part of the dream was me, on my knees, growling and barking myself hoarse, in the hope of being taken for a large and aggressive dog.

It was suggested to me by my therapist that it might be helpful not to spend quite so much time following the news. I didn't have to read everything, she said; most of the time it was enough just to glance at the headline. Though I took her basic point about duration of exposure, it was the headlines themselves that were the proximal cause of my distress.

This was a time in which payloads of apocalyptic force were delivered to the lock screen of my phone. It was the end of 2016, the winter of an ignoble year, and the more or less hourly vibration in my pocket was a kind of post-traumatic thrum, a bracing for whatever fresh hell I was about to peer into. I had come to think of my phone as my eschatology handset, my streaming service of last things. The world would end neither with a bang nor with a whimper, but with a push notification—a buzzing I wasn't even sure I'd felt, but figured

I'd better check anyway, to see if it was real, and what it might portend.

My wife—a person of unfathomable resilience and practical wisdom, to whom such fugue states of panic and epochal despair were essentially foreign—advised me to leave my apocalyptic obsessions at the door. The vibes were bad enough out there in the world, on the airwaves and the timelines, without my channeling them into the home. I was not John of Patmos, and this was not some cave of island exile: this was a house, and people were trying to live in it.

My therapist, another wise and practically minded person, made a remark that stayed with me. She did not intend for me to take the remark as a suggestion, she said, but it bore pointing out that a lot of people, when they experienced the kinds of anxieties I was experiencing, threw themselves into their work.

She didn't intend it as a suggestion, but it could be argued that I took it as such. It could be argued that this very book is the product of the work into which I subsequently threw myself. It is both a privilege and a curse of being a writer that throwing yourself into your work so often involves immersing yourself deeper into the exact anxieties and obsessions other people throw themselves into their work to avoid. I don't mean to suggest that this book began as some kind of therapeutic enterprise, but neither did it arise out of some sharp and rational focus of inquiry. The truth is that it arose out of a much stranger and more perverse motivation. I was anxious about the apocalyptic tenor of our time, it is true, but I was also *intrigued.* These were dark days, no question, but they were also interesting ones: wildly and inexorably interesting. I was drawn toward the thing that frightened me, the thing that threatened to tear everything apart, myself included.

Often, when I thought about this perverse motivation of mine, I thought about the narrator of James Joyce's story "The Sisters," who remembers as a boy being both repelled and fascinated by an elderly priest, paralyzed and dying in the wake of a stroke. Every night, he would repeat quietly to himself the word *paralysis.* "It filled me with fear," he says, "and yet I longed to be nearer to it and to look upon its deadly work."

I wanted to be near to the idea of the apocalypse, to look upon what evidence of its deadly work could be found in the present: not in the form of numbers or projections, which are nowadays mostly how it's revealed to us, but rather in the form of places—landscapes both real and imaginary where the end of the world could be glimpsed. And so this book is in some sense the outcome of a series of perverse pilgrimages, to those places where the shadows of the future fall most darkly across the present.

*Pilgrimage.* Why do I insist on such a weirdly religious, even self-aggrandizing usage? Because I was looking for something in these places, for some kind of enlightenment or edification or even solace. Needless to say I found these only fleetingly, but perhaps the wisdom is to be found in the looking, or so it suits me to believe. And all of the places that I encountered in the year or thereabouts I spent traveling for this book seemed to me to be charged with special significance, with the potential to reveal something crucial about the strange and nervous time, the hysterical days, in which I found myself to be living.

And there is this: a pilgrim is someone who travels in search of some earthly manifestation of their faith. If I could be said to have had a faith in those days, it was anxiety—the faith in the uncertainty and darkness of the future. And it was in search of manifestations of that anxiety that I set out. Because I wanted to look it in the face, this future-dread, to see what might be learned from it, what might be gained for life in the present.

Into what, then, did I throw myself, and where exactly did I land? That much is easy enough to say. I sought out places, ideas, phenomena, that seemed to me especially charged with these anxious energies. The first of these was my own home: at my desk, on the couch, peering at my phone in bed, I received at my leisure signs and portents of our ongoing apocalypse. Subjectively speaking, the apocalypse begins at home, which is one of the reasons why this book begins there, with my unwisely voracious consumption of YouTube videos about preparing for the collapse of civilization. Soon enough this interest in planning for and protecting oneself and one's family against the apocalypse took me farther afield, to the prairies of South Dakota, where I visited a former army munitions facility in the process of being converted into a "survival shelter community," a vast network of bunkers where those in a position to afford it could hide out in the aftermath of a cataclysmic event. I traveled to New Zealand, a country whose remoteness and stability have led to its reputation as a favored retreat of billionaires anticipating systemic collapse. And I attended in Los Angeles a conference on Mars colonization, an idea predicated on the conviction that we needed a backup planet on which our species might outlive a doomed Earth. In pursuit of a more psychological form of preparation, I went on a wilderness retreat to the Scottish Highlands, to a place ravaged by the twin forces of colonialism and industrialization, in the company of a group of people who shared my own anxieties about the future. And because I wanted to see what the end of the world—or, more accurately, *an* end of the world—might look like, I traveled to Ukraine, to visit the Chernobyl Exclusion Zone. In all of these places, the apocalypse was revealed to me in different forms: the cultural, the political, the scientific, the personal.

This book is about the idea of the apocalypse, but it is also about the reality of anxiety. In this sense, everything in these pages exists as a metaphor for a psychological state. Everything reflects an intimate crisis and an effort at resolving it. I went out into the world because I was interested in the world, but I was interested in the world because I was preoccupied with myself.

A final disclaimer: though this book might seem to be about the future, its true concern is the present moment. I offer no visions of what the future might be like—partly because I claim no authority from which to do so, but mostly because the future interests me only as a lens through which to view our own time: its terrors, its neuroses, its strange fevers. Either we are alive in the last days or we are not, but the inarguable thing in any case, the interesting thing, is that we are alive.

# 2

# PREPARATIONS

While reading the news, while submitting to the oblivion of my Twitter feed, I had taken to muttering under my breath, in half-conscious tribute to the narrator of Joyce's "The Sisters," the word *collapse.* It had a dark glamour to it, this word, and in its repetition there was a stern and oceanic comfort, like a perverse mantra. I thought about it a lot, this idea of collapse: what form it would take, what it might mean to live through it. It was, I knew, not a salutary thing to be spending so much time considering, even within the overall context of my generally anxious outlook.

For close to a year, my online homepage was set to r/collapse, a subreddit entirely devoted to news links and discussions pertaining to civilizational collapse and adjacent concerns. I would open my browser and would immediately be greeted by a crowd-curated selection of signs and portents, apocalyptic apocrypha. Black snow falling on Siberia. New strains of antibiotic-resistant bacteria. An iceberg twice the size of New York City that was breaking off from Antarctica.

*Business Insider*'s top ten major cities that could be unlivable within eighty years. Et cetera, et cetera, et cetera.

Even when I didn't click the links—which often enough I did not, for fear that what I gained in knowledge I would lose in sanity—my online existence was saturated in a sense of end-time urgency.

It would have been healthier, of course, not to mention more useful, to attempt to effect some small good in the world, or to challenge my energies toward some positive goal, but this did not seem to be how I was wired. Avoiding these more sensible options, I set out toward the darkness itself.

"Set out" is maybe not quite the right language here, as it connotes a certain degree of resolve, as though I were some kind of Kerouac-type figure, packing up his bag and striking out for adventure, which could not be further from reality. It would be more accurate, in fact, to speak of wandering or drifting toward the darkness, or even of loitering with intent at its margins. Although again, "intent" is potentially misleading in its own right, in that it inevitably gives the impression of my having literal intentions, which was hardly more true than my having resolve; and so perhaps it might be best to abandon these ambulatory metaphors altogether and proceed directly to the banal truth of the matter, which is that I was spending a lot of time on the Internet reading stuff about the apocalypse.

My obsession tended at first toward the sensationalistic, even within the general context of the end of the world. For a time, I pursued a leisurely and more or less abstract interest in the pursuit known as "prepping," a subculture made up, as far as I could see, pretty well exclusively of white American men who were convinced that the entire world was on the verge of a vast systemic rupture and were obsessively invested in making sufficient preparations ("preps") for such scenarios. It was all there in strange microcosm: the frontier mythos of freedom and self-sufficiency, the overwrought performance of masculinity that utterly failed to conceal the cringing terror from which it proceeded, the sad and fetishistic relationship to consumer goods, the hatred and mistrust of outsiders. Lurking on the forums and blogs and Facebook groups of these preppers—reading their literature and even listening to the occasional podcast—I came to see their movement as a hysterical symptom of America itself.

I punched in a great many hours on YouTube, hours that will never be returned to me, watching videos in which guys named Brandon or Kyle or Brent talked their viewers through the contents of their "bug-out bags," knapsacks containing the items they personally considered essential in any scenario whereby they needed to head out into the wilderness and fend for themselves.

It was generally projected that the worst effects of such collapse scenarios—nuclear attacks, mass civil unrest, viral pandemics, meteor impacts, so forth—would be focused on urban environments, and so the thinking among preppers seemed to be that the thing to do in a SHTF ("shit hits the fan") situation where you had to leave your home was to "bug out"—to head out into the relative safety of the wilderness, away from people. ("Bugging in," by contrast, was the preferred, albeit less exciting, option whereby you battened down the hatches and secured your current location: essentially a highly militarized version of not going out of the house, a form of apocalyptic preparedness that seemed more commensurate to my own skills and temperament.)

It struck me that these bug-out bag videos bore a strong family resemblance to the more mainstream YouTube phenomenon of "haul videos," in which a person, usually a young woman, laid out for her viewers the results of a recent shopping trip. The bug-out bag video was a kind of apocalyptic variation of this display of consumerist achievement. The revealed items were typically such things as hunting knives, first aid kits, head-mounted flashlights, extra batteries, multi-tools, crowbars, face masks, compasses, whistles, Kevlar socks, space blankets, military-grade cordage, mini water filtration systems, long-shelf-life emergency food ration bars, wet wipes, thermos flasks, camouflage-patterned duct tape, tea lights, cotton balls, Vaseline, and high-quality sunglasses.

In fact a minor tributary obsession of mine, flowing inexorably downward into the larger obsession with preppers and their consumer habits, was an interest in the specific style of sunglasses worn by these men: they seemed to overwhelmingly favor wraparound shades, a preference that was, as far as I could gather, more or less universal among right-wingers as a group. Footage of alt-right gatherings, Twitter avatars of libertarians, images of furious and red-faced men at Trump rallies: in all of these cultural artifacts, I noted the presence of this excessively curved and ovoid style of eyewear; but if there was some inherent connection between the wearing of Oakley-brand shades and the holding of extreme reactionary views—the staunch opposition to the role of the state in the structuring of society, the belief that personal liberty meant freedom from taxation, the conviction that white heterosexual males were in fact the last victims of societally sanctioned discrimination—I could not, for all my efforts, come up with any theory, either serious or frivolous, as to what it might consist of.

Around this time, I read a handful of practical guidebooks on preparing for the end of the world. The spirit in which I was reading them was never entirely clear to me. I was not completely sure of whether I was reading them out of abstract interest in prepping as a sociocultural phenomenon or out of a genuine desire to access esoteric knowledge of the coming collapse, of how to negotiate it when it eventually arrived. This uncertainty about my relationship to these books, and the fears they represented, was intensified at one point by my coming across a guidebook to post-collapse survival whose author bore the same name as myself.

The near-hysterically search-engine-optimized title of this book was *DIY Survival Hacks! Survival Guide for Beginners: How to Survive a Disaster by Using Easy Household DIY Techniques.* It seemed subpar even by the fairly lenient standards of the survivalist genre, apparently having been written in extreme haste against the looming deadline of the apocalypse itself. But the fact that its author had exactly the same name as me lent it a frisson of the uncanny, as though it should be received as a warning from the future. (That this other Mark O'Connell had also written a series of guidebooks on the interpretation of signs and symbols only served to intensify my sense that some barely encrypted omen was being retailed to me through Amazon.) Seeing my name on this book's cover—with its neat arrangement of tins, torches, first aid kits, walkie-talkies, candles, and bottled water—seemed to cause a small but irreparable rupture in the thin membrane of irony that had previously separated this subject from my own nerve endings.

The clearest delineation of the prepper mind-set I came across was a book called *How to Survive the End of the World as We Know It: Tactics, Techniques, and Technologies for Uncertain Times.* Its cover announced it as an "international bestseller," and its author, who was apparently a former US Army intelligence officer, went by the intriguingly weird name of James Wesley, Rawles. The comma, I gleaned from a Q and A on his website, was there to denote the distinction between his Christian name, which he felt to be his alone, and his family name, which was "the common property of all those that share the Rawles bloodline, and our wives." (The grammatical ambiguity of this sentence was such that I had to give Rawles the benefit of the doubt that it was the family name being shared, rather than the wives.) Rawles was a highly visible figure within the prepper movement: he ran a popular survivalist blog, was the author of a series of speculative dystopian novels about postapocalyptic survival—some of which were, not very surprisingly, set against the backdrop of a struggle against a global caliphate—and had founded a movement known as the American Redoubt, which advocated for a migration of like-minded conservative Christians and Jews (but apparently not

Muslims) to a sparsely populated region of the northwestern United States in order to prep for the twilight of civilization.

His *How to Survive* book portrayed an America teetering on the verge of collapse, in which the vast majority of the population was reliant for its food supplies on a tiny number of people and an increasingly complex distribution network. Any kind of major disaster, Rawles insisted—a mass outbreak of contagious disease, a nuclear attack, an economic collapse—could easily lead to people deciding not to go to work in the morning, and the shelves of Walmart consequently not being stacked, the delivery trucks staying off the roads.

"Crops will rot in the fields and orchards," he wrote, "because there will be nobody to pick them, or transport them, or magically bake them into Pop-Tarts, or stock them on your supermarket shelf. The Big Machine will be broken."

The average American family, he pointed out, had less than a week's worth of food in its possession at any one time. He conjured a scenario whereby the supply chain breaks down, and all of a sudden you've got hordes of "Joe Six Packs"—Rawles's contemptuous term for the unprepared suburban paterfamilias—striking out into the collapsing world in search of food and other supplies for their families. It's hard to ignore the distinct note of relish in the projection of this scenario, whereby these Joe Six Packs are suddenly shaken from their long dream of security and comfort and awoken to the harsh reality of TEOTWAWKI (The End of the World as We Know It): "The power grid is down, his job is history, the toilet doesn't flush, and water no longer magically comes cascading from the tap. His wife and kids are panicky. The supermarket shelves have been stripped bare. There are riots beginning in his city. The local service stations have run out of gas. The banks have closed. Now he is suddenly desperate."

The result of this situation, for Rawles, is the collapse not just of the nation-state and the economy, but of civilization itself, of a system of relations between individuals that was always, in this view, extremely delicate to begin with, always based on a necessary mass delusion as to the true nature of the world. You're looking at widespread looting, people stealing food and supplies from their own neighbors, mass violence, the total breakdown of law and order. At first, the chaos will be centered on cities, with desperate urbanites targeting restaurants and shops. But then, as food sources grow scarcer closer to home, these people will inevitably venture out into the provinces in search of sustenance for their families. Many of these people will form armed gangs, he predicts, running their vehicles on siphoned gas. These looters will finally come to grief as a result of the flu or of lead poisoning, but not, he warns, before causing a great deal of destruction.

This is a prediction of the future that could be offered only by someone who was never fully convinced by the idea of society in the first place. This seemed to me to be implicit in everything I learned about the preppers, and in everything Rawles wanted to impart to his readers. What he was offering was, in this sense, not so much a prediction of the future as a deeply political interpretation of the present. One passage in particular seemed to me to reveal the movement's total ideological abjection. "There is just a thin veneer of civilization on our society," writes Rawles. "What is underneath is not pretty, and it does not take much to peel away that veneer. You take your average urbanite or suburbanite and get him excessively cold, wet, tired, hungry, and/or thirsty and take away his television, beer, drugs, and other pacifiers, and you will soon see the savage within."

Apart from the extent to which it indicated Rawles's complete lack of investment in society itself, the introduction of the figure of the savage here was a lot more revealing than he presumably intended it to be. It had always seemed clear to me that, as a group, preppers were involved in the ongoing maintenance of a shared escapist fantasy about the return to an imagined version of the American frontier—to an ideal of the rugged and self-reliant white man, providing for himself and his family, surviving against the odds in a hostile wilderness. But what the use of the word *savage* made explicit here, I thought, was the extent to which this reactionary fever dream arose not out of any real understanding of the present or the future, but rather out of the historical trauma of America's originary apocalypse: the dehumanization and near-annihilation of indigenous peoples and their cultures.

And when preppers like Rawles invoked the specter of the savage, what they were doing was setting up a divide between themselves—as carriers of the flame of civilization, as heirs of the frontier spirit—and those who would immediately revert to a state of nature, the wild natives of the post-collapse world. And in setting up that divide, what they were further doing, whether they understood it or not, was creating the necessary conditions for a return to the cleansing violence of the nation's colonial past.

The most absurdly overdetermined manifestation of this energy was a video I watched on YouTube entitled "Top 10 Sheep Dog Gangs That Will Form After the Collapse," posted in August 2017 to the "Reality Survival and Prepping" channel. The video focused on a scenario that was obsessively fetishized among preppers: what they referred to as WROL, or "Without Rule of Law."

Sitting in a cramped room with the shades drawn, against the lurid backdrop of one of those "Don't Tread on Me" flags with the coiled rattlesnakes, a man named J. J. Johnson, who looked to be in his late thirties, laid out his understanding of American society and his vision of how, in the chaos that would immediately follow the coming collapse, certain elements of that society would coalesce to form a bulwark against savagery and lawlessness.

At present, in his view, there were two separate and irreconcilable Americas. There was urban America, which was densely populated and "controlled" by the Democrats, and there was rural America, where people went to church and were enthusiastic about the recreational use of firearms. He went on to speak of "good guys," and although he did not explicitly mention "bad guys," it was fairly clear that his division of America into two implied a "good" America and a "bad" one. America, he said, had a lot of good guys—a lot of good guys with guns—who would not put up with "a lot of lawlessness." (In an aside, he referred his viewers to a previous video in which he discussed how, in an SHTF scenario, looters and "brigands" would be killed first.)

Johnson's overall thesis, it quickly became clear, was that

certain preexisting social groups would emerge as de facto law enforcement during the breakdown of civilization, maintaining—by violence if necessary—the sanctity of private property and the safety of "good" American families. These groups he referred to as "sheep dog gangs." ("And when I say 'gang,'" he hastened to clarify, "I'm meaning it just as in a group of people. That's really all this is. It doesn't mean they're going to behave like an urban-style city gang.") Examples of existing groups that would form these postapocalyptic "sheep dog gangs"—or "posses," as he referred to them at one point—were men's organizations, homeowner associations, veterans' groups, and local chapters of the Rotary Club.

For all that I relished the absurdity of Rotarians coming together to defend civilization, I was also capable of recognizing crypto-fascism when I saw it, and this seemed to me as good an example as I'd encountered in the wild. This vision of God-fearing members of the business community and war veterans coming together to defend themselves against an onslaught of urban looters and general lawlessness was plainly a fantasy of purgation, focused on the violent elimination of "bad" elements of American society.

In fact, you couldn't even properly call it crypto-fascism: it was really just good old-fashioned original-style fascism. It didn't seem necessary either to do a lot of racial decoding when it came to all the talk of "urban" versus "rural" America, of "city-style" gangs versus homeowner association posses. The whole appeal of the apocalypse, for this J. J. Johnson, seemed to be that it gave him a pretext for this kind of Klan-style fantasy.

—

The idea of collapse speaks, on some primal level, to a reactionary sensibility—a sensibility in which the world is always necessarily in an advanced state of degeneration, having fallen from a prelapsarian wholeness and integrity. (Feminism, political correctness, the supine attitude of the left toward Islam, and so on: the structure of Western civilization is, in the reactionary view, always being eaten away from within.)

Prepping is rooted in the apprehension of an all-consuming decadence. Society has become weak, excessively reliant on systems of distribution and control whose very vastness and complexity renders them hopelessly vulnerable. The city is the source of this decadence. Preppers don't trust cities, or those who make their lives in them. All those people living at a remove from the production of food, completely reliant on those vast and fragile systems, of distribution and waste collection, those heaving masses of humanity, incorrigibly plural and various. And what this suspicion amounts to is a suspicion of modernity itself.

Take the show *Doomsday Preppers,* which aired on the National Geographic Channel from 2011 to 2014. On the surface level, this was a reality television program about American men making elaborate preparations for the collapse of civilization: building fortresses, bunkers, remote rural hideouts; stockpiling weapons, tools, foodstuffs, and other postapocalyptic essentials. But you don't have to sit through very many episodes on YouTube to understand the extent to which the show is in fact a reality TV psychodrama about masculinity in crisis.

The protagonists of the show, typically middle-class rural white men—not especially wealthy or highly educated, but comfortable enough to invest significant proportions of their income on fantasies of rugged self-sufficiency in the wake of a great civilizational crackup—are uniformly obsessed with purifying their lives of dependence on others. These men's critique of modernity, such as it is, is a critique of the extent to which the individual has become weakened and compromised by such dependency. "I was raised to not rely on anybody," as one prepper puts it, in a more or less representative statement of the movement's politics. "Don't rely on your government, don't rely on your neighbors. You count on yourself first."

The show is grindingly repetitive—once you've seen one middle-aged American white guy pursue his elaborate fantasy of individualism and self-sufficiency, you've pretty much seen them all—but there is a bleak comedy to the spectacle of these men performing, and imposing on those around them, their understanding of what it means to live in the "real world." They are, almost to a man, deeply invested in the notion of their own pragmatic approach to life, and to the future. These preppers are often surrounded by people—usually women—who are less driven by the desire for independence, less maniacally certain about the advent of apocalypse, and who must therefore be taken in hand, initiated into the ways of the prepper.

In one episode, we meet Brian Murdock, a Massachusetts real estate broker and devout Christian. Brian is the consummate prepper, in the sense that every aspect of his life appears to be subservient to the overall project of readying himself for the collapse of civilization. (The particular scenario in Brian's case is a third world war, arising out of a nuclear attack by the United States on Iran, avenged by a counterattack on Israel. "I know this with every fiber of my being," he says, leaning against the wooden porch of his colonial home, lemon polo shirt tucked into the waistband of his checked board shorts. "One-third of the earth will perish.") His decision to marry, he says, was taken after learning that the chances of surviving a catastrophic event greatly increase when you have a partner who is invested in your survival.

And so, in an apparent effort to reverse engineer a normal human relationship from the premise of his own self-interested survivalism, he joins a dating website and meets a young woman from Colombia named Tatiana. Having made a couple of trips down there to get to know her, he arranges for her to come to America as his fiancée. He is particularly drawn to Colombia as a wife-sourcing location because he has heard that their way of life is very simple, and that they have a culture of "gratitude" and "respect." Brian doesn't mention feminism, but it seems implicit, as we watch his intended young bride loading the dishwasher after her first meal in America, that traditional gender roles are an important part of his vision for postapocalyptic survival.

"I believe that the blessing of marriage," he says, "the *covenant* of marriage, is very central to prepping."

Having arrived in the United States, Tatiana is taken aback to be told about Brian's nuclear war plan to retreat to a fifty-acre property seven hours' drive away. Before she is even allowed to unpack her suitcase, she is made to prepare a bug-out bag. None of this—prepping, bugging out, nuclear contingency plans—had come up in their conversations before she'd come to America, and it all seems strange and confusing to her. "When Brian told me he was a prepper," she says, "I thought he was crazy. In Colombia we don't do it. We don't know about saving food for bad times, you know, because there are bad times all the time already." But her husband is wise, she says, and she is committed to one day becoming the perfect prepper wife.

The show overwhelmingly presents women as naive and in need of tutoring in the ways of prepping. Wives are skeptical, concerned with the frivolities of everyday life, but invariably come round in the end to the necessity of regular drills, firearm training, and so forth. Daughters are taught to fear "marauders," hungry men with rage in their hearts and lust in their eyes. The family becomes a kind of fortress against the dangers of the world, the father a figure of feudal paternalism, offering protection through the skillful dispensing of violence, masculine know-how, and ingenuity. There is a fetishizing of older ways of being a family, of how things were imagined to be before the advent of feminism and other corrosive social forces.

Preppers are not preparing for their fears: they are preparing for their fantasies. The collapse of civilization means a return to modes of masculinity our culture no longer has much use for, to a world in which a man who can build a toilet from scratch—or protect his wife and children from intruders using a crossbow, or field dress a deer—is quickly promoted to a new elite. The apocalypse, whatever form it takes, will mean misery and death for most human beings, but for the prepared, it will mean a return to first principles, to a world in which men are men. Especially if they are white.

The racial dimension of prepper anxieties and fantasies is impossible to ignore. It's there in the distrust of cities and their inhabitants. It's there in the clarification, in that YouTube video, of the distinction between "sheep dog gangs" and "urban-style city gangs." And it's there in all the excited discussion of looting and "lawlessness," and the things that might have to be done to keep them from getting out of hand. When preppers talk about "civil unrest," as they frequently do—it is, after all, one of the primary methods by which the shit is expected to hit the fan—they invariably seem to be referring to black people and their reactions to systemic racism and violence. In 2015, when a grand jury declined to bring charges against six police officers involved in the death in custody of Freddie Gray, the city of Baltimore erupted in riots, and these riots were viewed by many preppers as a harbinger of precisely the kind (and precisely the color) of WROL scenarios they are preparing for.

Example: in the immediate aftermath of the civil unrest in Baltimore, a website called *The Prepper Journal* posted an article whose title posed the presumably rhetorical question "What Would a WROL World Look Like?" If the photograph that appeared on top of the post could be taken as its own kind of answer, a WROL world would look like a group of young black men, hooded and masked, jumping on the roof of a police cruiser. The article itself never mentioned Baltimore or Freddie Gray or the Black Lives Matter movement, but the photograph's provenance was obvious. (The side of the police car, for one thing, bore the words "Baltimore Police.") Farther down the page was another lawlessness-themed photo, depicting a hooded black man in the act of hurling an unseen object, behind him a parked car engulfed in flames. A reverse Google image search confirmed my suspicion that the photo was taken during the London riots of 2011.

Two things immediately struck me about this article, and the photos the author had chosen to illustrate it. The first was the more or less explicit association of lawlessness with young men of color. The perceived savage population here—the sector of society whose natural inclination toward violence and chaos would be given immediate expression in the event of a systemic breakdown—was emphatically nonwhite, emphatically urban. And the second was that the context for both photographs, Baltimore in 2015 and London in 2011, was widespread grief and rage about the death of a young black man in police custody. The people who were being made to stand for a world without rule of law, in other words, were those who understood most intimately and urgently what it meant to live without the protection of the state, to know that the law had never been intended to protect them in the first place.

The failure to acknowledge, or even to perceive, the lengthening shadow of the vast dramatic irony that attended this whole matter—namely that it was precisely society's most marginalized and oppressed people who truly understood what it might mean to live in a postapocalyptic world, and who were therefore most fully prepared—seemed to me to indicate a total moral incapacity.

This moral incapacity was something I'd recognized in Rawles's book, too, most memorably in a story he'd related about his time as an army intelligence officer in Iraq. One of the things he observed on the job, he says, was that in situations of structural collapse—such as, presumably, the collapse of Iraqi society that had resulted from his own government's illegal invasion in 2003—it was always refugees who suffered the most in these situations. And the moral lesson he drew from this observation, the idea that he took home with him, did not, despite the Christianity he claims as absolutely central to his identity, involve any kind of imperative to alleviate the suffering he'd witnessed, but rather a steadfast personal commitment: "I vowed," he writes, "never to become a refugee."

And this void of empathy seemed to me by no means incidental to the prepper movement, but rather a constitutive element of the entire project, the moral void around which it was structured. To be a prepper was to do everything one could do to avoid being one of the sufferers oneself, while contributing nothing to the prevention or alleviation of suffering in others.

Given their obsession with the prospect of collapsing distribution networks, and with the consequent need for self-sufficiency and self-reliance, the preppers' relentless fetishization of consumer goods seemed deeply ironic to me, if also basically understandable. The forums were filled with endless discussions on, say, which thermos flask or which flashlight would be the trustiest option in a SHTF scenario, and a small but apparently thriving economy seemed to have grown up around the demand for various gadgets and comestibles catering to the postapocalyptic survival fantasies of American men.

I came across one company called NuManna, named in reference to manna, the foodstuff the god of the Old Testament had provided for the Israelites during the time of their wandering after the Exodus. The company marketed gigantic buckets of freeze-dried powdered foodstuffs with a shelf life of a quarter of a century, whose varieties included, but were by no means limited to, oatmeal, hearty beans and beef, cheddar broccoli soup, and pasta primavera mix with freeze-dried chicken chunks.

In the Testimonials section of NuManna's website, I read a brief blurb from a customer named Reagan B., which seemed to me an unwitting encapsulation of the absurdity of the entire apocalypse preparedness project. "This stuff is awesome," wrote Reagan. "My wife has been away for a while so I ate NuManna while she was gone. It was simple and everything I had was really good. I wish NuManna was around when I bought a bunch of bulk food in the past from the Mormons. I don't want to have all these ingredients and put them together. NuManna was simple and great tasting. I gave away all my other bulk food."

At first this comment seemed purely and unimprovably comic in its conjuring of a character who, for all his determination to be adequately prepared for the collapse of civilization due to nuclear war or the impact of a massive asteroid, was also the type of man for whom not having his wife around to cook dinner—which seemed to me to be at worst a Domino's Pizza situation—forced him to crack open his apocalyptic food stash. (Equally bewildering, equally wonderful, was his purchasing food in bulk only to conclude that he lacked the stomach for the labor of assembling all these ingredients into meals.)

But on further consideration, the comedy gave way to something darker and more poignant: the idea of a man whose obsession with preparing for the end times had been so alienating and painful to his wife that she had left him, thereby bringing about a kind of personal apocalypse whose outcome was this helpless, fearful, obsessive person subsisting off a supply of flavored protein sludge he had amassed for the literal end of the world.

And this was a man with whom I identified. Of course I identified with him: I'd all but invented him. (I was his hypocrite reader, I thought: *mon semblable, mon frère!*) He was an outlandish avatar of my own anxieties and meta-anxieties—my anxieties about the damage my ongoing state of anxiety might be causing.

Because that was the thing about preppers: they were easily ridiculed, and their politics made it tempting to outright disdain them, but at an instinctual level I felt that I understood where they were coming from. Though I didn't share their manic insistence on preparing for the collapse of civilization, I knew the distributed matrix of unease from which the certainty of that collapse grew. I, too, with my pessimism, my intimate imagination of the world's unraveling, had driven my own wife, if not to despair itself, then to somewhere in its vast and crumbling exurbs.

The possibility has from time to time occurred to me that my contempt for preppers is exacerbated by a suspicion that I might not be as different from them as I like to imagine. The politics of the movement—the cringing fear of the poor, the dark-skinned, the feminine, the other—are reprehensible to me, but their sense of the fragility of the systems by which we live is, in the end, hard to dismiss as entirely paranoid, entirely illogical.

In the nontrivial number of conversations I had in that time that centered around anxieties of civilizational collapse, it quickly became clear to me that I was not alone among my friends and acquaintances in the suspicion that a hazily delineated catastrophe was taking shape on the horizon. More than a few friends informed me that they had given some thought to the possibility of stocking up on supplies for some kind of apocalyptic scenario, though for most of them this never went much further than idle consideration. Either they didn't have the space to build bunkers, or they were too lazy, or—and this was by far the most common reason—they concluded that if civilization were to actually collapse, they would much rather be dead than try to survive whatever cataclysm might be in store, because who in their right mind would really want to survive a nuclear holocaust or asteroid impact event anyway?

I myself find that even reading the words "pasta primavera mix with freeze-dried chicken chunks" is extremely helpful in clarifying my own stance on the question. If the choice on offer is between pasta primavera mix with freeze-dried chicken chunks and being among the first wave of deaths in the apocalypse, I hereby enthusiastically place my order for oblivion.

Because I was spending so much time in those days thinking about the prospect of collapse, and watching prepper YouTube videos, the topic came up frequently enough in conversation. People would ask what I was working on, and I would say I was thinking of writing something about people preparing for the end of the world, and people—friends, acquaintances, people I had just met—would tell me about their own such anxieties, or those of people they knew.

One friend of mine, who rented a room in a large house in London owned by a very wealthy friend of his, told me that this friend's mother, who was the eccentric heiress of one of the first great American fortunes, would occasionally call around to the house and hold forth on the near certainty of imminent systemic collapse, and insist that my friend start buying large quantities of canned foods, even going so far as to offer to pay for the construction of a small bunker in the back garden, an offer which—for all the ironic temptations of installing an apocalypse shelter paid for by the same fortune that had contributed greatly to the construction of Manhattan in the nineteenth century—was in the end diplomatically rebuffed.

Then one day I had lunch with a friend of mine, Sarah, who worked in publishing. I knew that she shared some of my apocalyptic fixations, but I had not been aware of the depth and seriousness of her obsession with the end of the world. Under her bed, she said, she kept a large backpack, ready to be hauled out at a moment's notice. Inside it was a tent, and a miniature camping stove, and a selection of knives, and chlorine tablets for water purification purposes. There was a compass in there, too, and actual paper maps, which would remain useful long after the phone networks went down. This backpack of hers—her "go-bag," as she called it—she had taken out to the wilderness on solo excursions that seemed to be somewhere between camping trips and emergency drills. Sarah, it turned out, was an honest-to-God prepper.

She claimed to find the whole thing vaguely embarrassing, but it seemed clear that there was also some measure of pride. The whole civilizational collapse scenario was appealing, she said, in that you would be tested to the limits of your resourcefulness, resilience, and self-sufficiency. You would, in the absence of any kind of societal structure, quickly learn what you were made of. Wasn't there something exciting, she asked, about that prospect?

I myself had no interest, I told her, in finding out what I was made of. My suspicion was that it was not first-rate material. Whatever form the apocalypse came in, I would almost certainly be in the first wave of deaths. We both laughed, but I think she knew that I was not entirely joking.

I said it sounded like her thinking was broadly in line with the prepper movement as a whole, but that she was just more honest about the extent to which she was driven as much by fantasy as by fear. There was, she said, undoubtedly an element of wish fulfillment, but in a way that was very different for her as a woman than it was for the majority of preppers who were men. Theirs was a fantasy of return to patriarchal norms, to a prefeminist dispensation that would be reestablished after the breakdown of civilization. But women, said Sarah, were already halfway to a dystopia. If she got raped tomorrow, for instance, she was by no means confident that she would go to the cops about it. I took her point to be that civilization was a relative concept to begin with, and that its collapse could seem to be more or less under way, depending on where you were standing.

And then she said something that I had not previously heard from any prepper. She was aware, she said, of a desire for final knowledge. To think that it might be to us, in our time, that the end of the story would be revealed: Was there not, she said, some comfort in that, some satisfaction?

I didn't know how to answer the question. In an abstract sense, a cultural sense, I understood this as some part of the psychology of apocalypse. But as an individual, as a parent, I wanted the world to live on after me. This, I said, was at any rate my assumption about myself. But perhaps the reasons for my interest in the end of the world were more complicated than I was prepared to acknowledge. Perhaps my terrors and my desires were more intimately related than I knew.

# 3

# LUXURY SURVIVAL

The week I traveled to the Black Hills of South Dakota to see the place from which humanity would supposedly be reborn after the coming tribulations, there happened to be a lot of talk about nuclear war. The UN had announced sanctions against North Korea, and North Korea had vowed to take physical action against such sanctions, and America, in the person of a president who was at that point vacationing at one of his many eponymous luxury golf resorts, advised that if they so much as lifted a finger they would be met with "fire and fury like the world has never seen." According to *The Wall Street Journal*, analysts were trying to guess what would happen to the markets in the event of all-out nuclear war between the United States and North Korea. (The answer seemed to be that you would likely see some flattening of yield curves due to lower risk appetites, but that from a financial perspective a nuclear apocalypse wouldn't exactly be the end of the world.)

The apocalypse was trending. The memes were dank with foreboding, and the presiding mood was one of half-ironic Cold War nostalgia mixed with sincere eschatological unease. It seemed as good a time as any to visit a place for sitting out the end times. My obsessive consumption of prepper videos had opened out onto a broader vista of apocalyptic preparedness, and to a lucrative niche of the real estate sector catering to individuals of means who wanted a place to retreat to when the shit truly hit the fan.

I'd made arrangements to meet with one Robert Vicino, a real estate impresario from San Diego who had lately acquired a vast tract of South Dakota ranch land. The property had once been an army munitions and maintenance facility, built during the Second World War for the storage and testing of bombs, and it contained 575 decommissioned weapons storage facilities, gigantic concrete and steel structures designed to withstand explosions of up to half a megaton. These Vicino intended to sell for twenty-five grand a pop to those Americans who cared to protect themselves and their families from a variety of possible end-time events—from nuclear war, certainly, but also from electromagnetic pulse attacks and gigantic solar flares and asteroid collisions and devastating outbreaks of viral contagion, and so on. He had set up shop in the Black Hills in the hope of drawing some customers from the multitude of bikers who had descended on South Dakota that week for the Sturgis Motorcycle Rally.

Vicino was among the most prominent and successful figures in the doomsday preparedness space, a real estate magnate for the end of days. His company specialized in the construction of massive underground shelters where high-net-worth individuals could weather the end of the world in the style and comfort to which they had become accustomed. The company was named Vivos, which is the Spanish word for *living.* (As in *los vivos*—as distinct, crucially, from *los muertos.*) Vivos claimed to operate several facilities across the United States, all in remote and undisclosed locations, far from likely nuclear targets, seismic fault lines, and large urban areas where outbreaks of contagion would be at their most catastrophically intense. They were advertising an "elite shelter" in Germany, too, a vast Soviet-era munitions bunker built into the bedrock beneath a mountain in Thuringia. Vivos's flagship location was beneath the cornfields of Indiana. It had been a government shelter during the Cold War and supposedly featured a luxuriously appointed dining area, a home entertainment theater, a medical center equipped for surgery and defibrillation, a pet kennel, and a hydroponic miniature farm for growing fresh fruit and vegetables. The development also apparently boasted the world's only private DNA vault—described by Vicino as "the next ark of humanity"—in which members could store their own genetic code, "for the preservation and potential restoration of the human race itself."

Vivos's new South Dakota location went by the name xPoint. Each of the bunkers, evenly spaced across eighteen square miles of prairie land, had an area of 2,200 square feet—significantly larger than my own (admittedly not very large) house. The place would, it was claimed, be home to somewhere between six thousand and ten thousand people and would become "the largest survival community on earth." It was pitched at a demographic somewhere between the super-wealthy clients for

Vivos's luxury underground shelters and the doomsday preppers who planned to survive the apocalypse through manly grit and YouTube know-how. It was the future domain, in other words, of the postapocalyptic petit bourgeoisie.

The place was, I read on the company's website, "strategically and centrally located in one of the safest areas of North America," at an altitude of about 3,800 feet and some one hundred miles from the nearest known military nuclear targets. "Vivos security team can spot anyone approaching the property from 3 miles away. Massive. Safe. Secure. Isolated. Private. Defensible. Off-Grid. Centrally located." (It was not intuitively clear to me how a place could be both isolated and centrally located, but then such claims were no more lavishly contradictory than the sort of thing you'd come across in even the most mundanely pre-apocalyptic of property listings. At any rate, if pretty much the entire rest of the world had perished—nuclear attacks, cannibalism, paroxysms of miscellaneous savagery—any settlement of living humans would have legitimate grounds to proclaim itself centrally located.)

Vivos was offering more than just the provision of ready-made bunkers, turnkey apocalypse solutions. It was offering a vision of a post-state future. When you bought into such a scheme, you tapped into a fever dream from the depths of the libertarian lizard-brain: a group of well-off and ideologically like-minded individuals sharing an autonomous space, heavily fortified against outsiders—the poor, the hungry, the desperate, the unprepared—and awaiting its moment to rebuild civilization from the ground up. What was being offered, as such, was a state stripped down to its bear right-wing essentials: a militarized security apparatus, engaged through contractual arrangement, for the protection of private wealth.

End-time real estate was an increasingly competitive racket. In Texas alone, you had two major purveyors of luxury apocalypse solutions: Rising S, a manufacturer of high-end customized shelters and bunkers, and Trident Lakes, a planned community north of Dallas featuring a range of variously lavish condominiums priced between half a million and two million dollars. On the Trident Lakes website, I read that in the event of a nuclear, chemical, or biological emergency, the properties would be sealed by automatic airlocks and blast doors, and that each would be connected via a network of tunnels to an underground community center featuring dry food storage, DNA vaults, fully equipped exercise rooms, a greenhouse, and meeting areas. The promotional blurb also promised such features as a retail district, an equestrian center and polo field, an eighteen-hole golf course, and a driving range.

This was a new entry into the apocalyptic imaginary: bankers and hedge-fund managers, tanned and relaxed, taking the collapse of civilization as an opportunity to spend some time on the links, while a heavily armed private police force roamed the perimeters in search of intruders. All of this was a logical extension of the gated community. It was a logical extension of capitalism itself.

And it brought to mind an image that had gone viral around that time: a photograph of three men obliviously golfing against the backdrop of an Oregon wildfire, a sheer wall of incandescently burning pine forest rising like a vision of the inferno itself above the impeccably maintained greens. It was like a Magritte painting in its surreal juxtaposition of irreconcilable realities. The first time I saw it on my Twitter feed, my reaction was one of almost vertiginous moral horror. It was almost too terrible, too bizarre, to assimilate. And then I kept seeing it, over and over, until my reaction became: *this again?*

My point is that it did not take long for me to become, in my own way, one of the golfers.

Waiting for a call from Vicino to arrange our meeting, I had nothing better to do than mooch around Hot Springs. It was Sunday, and the town was largely deserted, save for a steady procession of grizzled and leather-vested bikers passing at a respectful clip through Main Street en route to Sturgis, many of whom had hoisted Old Glories like ensigns from the sterns of their Harleys, flags so incommensurately massive as to provoke faint anxieties about wind-drag and possible capsizing. In the town itself, I noted the omnipresence of the same flag—on car bumpers, in store windows, on sundry items of clothing, billowing superb and regnant from otherwise unremarkable buildings—and was impressed by the melancholy strangeness of this insistent motif, which seemed to me a kind of obsessional warding off of its own meaninglessness.

In a café on Main Street, I sipped a coffee and scribbled in my notebook, before being driven away by a loose but resilient alliance of flies, who took turns in alighting on my forearms as

I wrote. I walked the bank of the river, giving a wide berth at one point to a yellow-striped snake as it hustled its way across the path and up a grassy slope, and then I followed on a whim a sign for the Fall River Pioneer Museum.

I was as it happened the museum's only visitor, and I found myself unnerved by the silence of the building, and even more so by the chipped and peeling mannequins stationed about the place, dressed in nineteenth-century apparel: long silk gloves, black gauze veils and bonnets. At the summit of the house, in a large, creaking room devoted to agriculture, I encountered an exhibit that caused my heart to momentarily falter: a pair of rampant Friesian calves, taxidermied in a freakish embrace, the hooves of their forelegs resting on each other's shoulders. According to the laminated card in front of them, they had been born "joined at the brisket." The sense of indeterminate augury I had felt since entering the museum rose to consciousness now in the presence of these unreal animals. In the Middle Ages, I remembered, conjoined births were an omen of ill times, and during periods of widespread struggle and turmoil their appearance was seen as an outright portent of apocalypse.

On my way out the door, the gnarled old man at the desk remarked that I'd gone through the museum pretty quick. I could have been picking him up wrong, but he seemed a little put out.

"Be sure to check out the iron lung in the garage out back," he said, in a manner both rote and wistful, and I assured him that I would, but I did not.

As I walked down the hill, my phone vibrated in my pocket. Vicino was out at the site and was ready when I was.

—

About ten minutes after turning off Route 18 onto the cracked interior roads of the ranch, I passed what was once the town of Fort Igloo, home to the hundreds of workers who moved there to take up jobs at the Black Hills Ordnance Depot, built in 1942 to service the army's increased wartime need for munitions testing and storage. Schools, a hospital, shops, houses, a church, a small theater: all abandoned now to the oblivious cows. Only once the hollow carapace of Fort Igloo began to recede in the rearview mirror did the landscape reveal the true depth of its uncanniness, because it was then that I saw the vaults. I noticed at first only three or four of these things: low, grass-covered protuberances, spaced a few hundred feet apart, their hexagonal concrete frontage jutting from the earth. The deeper into the ranch I drove, the more of these structures emerged from the landscape, until I realized that they were everywhere, hundreds of them, as far as I could see in every direction. It was an ethereal sight, both alien and ancient, like the remnants of a vast religious colony, a place built for the veneration of derelict gods.

I stopped the car and got out, took a couple of pictures with my phone. But the structures barely registered in the images, which reduced the scandalous vastness of the landscape to the level of the banal. The immense horizontality, the otherworldliness, could be properly viewed only with the naked eye.

I drove another couple of miles and came across a large, empty barn, beside which was a dark brown shipping container the size of a small house. On one side was a banner that read "xPoint: The Point in Time When Only the Prepared Will Survive." Parked next to it was the silver Lexus SUV I'd been told to look out for.

I walked up the steps into the shipping crate, and into a kitchenette. From a room in the back, a gigantic man in his early sixties emerged and ambled toward me, and immediately embroiled me in a painfully vigorous handshake. Robert Vicino was six feet eight inches and, according to his own most recent computations, 310 pounds. ("By no means a porker," he said, patting the expansive dome of his abdomen. "Just a big guy.") The crimson bulb of a nose, the pockmarked face, the neat gray goatee: before he even began to speak—which he quickly did, and never let up—he presented himself to me as a distinctly Mephistophelian figure.

Before long, we were in the Lexus, getting ready to head to the nearest town to get diesel for the generator. His seat leaned backward at an absurdly steep angle, Vicino removed a large wooden-backed hairbrush from a side compartment and began to groom, with firm and precisely rhythmical sweeps, first his beard and then his hair.

"This is a great car," he announced. "You guys got Lexuses in the UK?"

"We have them in Ireland, anyway," I said, a little more sharply than I'd intended. "Not me personally, but people have them."

"Best car I've ever owned. And I've owned Mercedes. I've owned Rolls-Royce."

Sitting in the back was Jin Zhengii, a twenty-three-year-old recent engineering graduate whom Vicino had hired as his intern. Jin didn't say much—partly on account of being Chinese and not having particularly good English, but mostly, I guessed, on account of just being the kind of person who didn't say much.

"I tell him, Jin, I'm like your American dad," said Vicino. "Right, Jin? He's a great kid. Great kid."

They'd eaten at a Chinese restaurant in Rapid City the previous night, and Vicino had gone out of his way to fix him up with their waitress.

"I figured she was at least an eight or a nine," he said. "Jin was like, in China this girl is a three, tops. Right, Jin?"

In the back seat, Jin made a display of indifference—a roll of the shoulders, a tilt of the head—confirming the girl's low rating in a Chinese context. Robert told him to get her Facebook up on his phone, then took the phone from him and started swiping through photos of her, presenting them for my approval.

"I mean come on, look at this," he said, showing me another photo. "A three? Does she look like a three to you?"

As we drove, I gazed out the window at Fort Igloo, or the ruins thereof, and I was struck by the sense that I was apprehending at once both the past and the future. Vicino mentioned as we passed that the place had been home to hundreds of families. The news anchor Tom Brokaw, he said, had grown up here after the war, a fact he seemed to take a proprietorial pleasure in presenting to me for inspection. A concrete staircase with metal banisters stood in the middle of a field, utterly alone, no trace remaining of the building that once presumably provided it with a context.

We reached the dismal little town of Edgemont, which had foundered badly in the years since the ordnance facility had closed. The streets were long and narrow and seemed entirely desolate of human life. There was a Laundromat, a low corrugated iron structure, called Loads of Fun Laundry. Outside a gas station, a cluster of bikers were standing around beside their Harleys, variously bedecked in patriotic insignia. They were, I noted, all wearing the ideologically appropriate wraparound-style sunglasses.

"I'm gonna go talk to these guys," said Vicino, as he piloted the Lexus suavely into the gas station's forecourt. "Jin, you wanna look after the diesel?"

He explained to me a little running joke he had with himself when it came to bikers. He'd approach them and ask them very civilly what they would do if he were to just kick over their bikes onto the street. Most recently, he'd aired it with a couple of motorcycle cops back in California. Without fail, he said, the reaction was one of disarmed amusement.

"The one cop was like, 'You'd hurt your foot is what'd happen.' "

Vicino was, among other things, a man who knew how to exercise his whiteness to its fullest extent. He was going to try out his joke as an opening gambit with these bikers here. They were, after all, pretty much right in his target demographic: these guys tended to be self-reliant types, he said, not big fans of government. And despite appearances, a lot of them were doctors, lawyers, professionals, retired folks with money to spend. (My own preconceptions about bikers as a predominantly working-class demographic had been undermined during a brief stop-off at Mount Rushmore the previous day en route from Rapid City. Standing on the viewing platform, gazing across at that absurd and yet somehow moving monument to American grandiosity, I happened to be positioned next to two burly and leather-jacketed bikers, whose conversation about their respective legal secretaries I could not help overhearing.)

He'd been sitting in a café in San Diego last year, he told me, when he'd received an email from a cattle farmer up in South Dakota, informing him about the vast tract of land on his ranch, its former munitions vaults, and how it might be a suitable property for his business to acquire. The plan came to him instantly, he said, the whole idea for xPoint: he was going to pay the rancher the sum of one dollar for the property, offering him a 50 percent cut of all future profits from the vaults, which he was going to sell for thirty-five grand a pop to people willing to fit them out to their own specs, and it was going to be the largest survival community on Earth. It was going to be a much more affordable proposition than his other survival communities: an apocalypse solution for consumers of more modest means. He'd already sold off fifty or so.

"I'm thinking I might put a titty bar in one of the bunkers," he said at one point. "Like a mud-wrestling lesbian kind of deal."

This was another running joke he had with himself, a line he used whenever people asked too many questions about his projects. Like the time in Indiana, when the lady at the paint shop had gotten curious about the sheer volume of paint he and his team were buying, about the frequency with which they kept coming back to the shop, and had point-blank asked them what the hell they were building out there.

"I told her we're building a mud-wrestling titty bar," he said. "That was the end of that conversation."

The sound of the door closing was like nothing I had ever heard, an overwhelmingly loud and deep detonation, the obliteration of the possibility of any sound but itself—so all-encompassing and absolute that it became almost a kind of silence. It lingered in the empty interior of the vault for what felt like perhaps three or four minutes, taking full command of the darkness. It was an apocalyptic sound, and I was both unnerved and exhilarated. The darkness, too, was absolute, an annihilation of the very concept of light. As I stood in the reverberating void, it struck me that the fear of darkness was not so much the fear of what might be out there, unseen and moving, but rather the solipsistic and childlike terror of there being in fact nothing out there at all, that a world unseen was a world that had ceased entirely to exist.

If I'm seeming to imply that I was having cool and abstract insights into human psychology there in the black vault, let me state again that my primary emotion was fear. I was, in fact, momentarily deserted by my faculties of reason and began to panic that I might never get out of this place. I was pretty sure that the bolt was on the outside of the door. What if Vicino was in fact a madman, a homicidal lunatic who had decided to immure me in here, like a poorly characterized antagonist in one of Poe's tales of terror? What if he'd decided I was going to sell him out, that I was likely to damage his business prospects by making him out in my book to look like a fool, or a charlatan, or the kind of Poe-esque madman who might murder an adversary by entombing him in a decommissioned weapons silo, and that his only recourse was to encrypt me alive somewhere in the lonesome Black Hills of South Dakota, where even if there was another human being for miles around, which there was not, they would never hear my cries for help? Or what if—and this struck me as a more likely scenario—he was suffering a massive cardiac arrest out there, perhaps brought on by the effort of heaving shut the reinforced metal door, and was right now keeling over face-first into the dirt? He was a vast man, even perhaps a giant, and such people were prone to early death from heart attacks—not to mention the presumably considerable stress of spending all your time thinking about the end of the world, of constantly envisioning asteroid collisions and government cover-ups and collapsing coastal shelves and total nuclear war. How long would it be before I was found? The fact that Jin was standing beside me was slightly reassuring in terms of the first scenario, but not at all in terms of the second.

The void filled with sunlight then, and as my eyes adjusted to the brightness I was able to make out Vicino's prodigious silhouette in the doorframe.

"How about that? Isn't that something?" he said cheerfully. I agreed that it was, and the slight quiver in my voice was engulfed and obscured by the rapid echoing of my words in the emptiness.

—

Later, Vicino told me of how he'd made his money in advertising back in the eighties. He'd basically pioneered what was known as "large inflatables." His hour of glory had arrived in 1983 when, to mark the fiftieth anniversary of the release of the original *King Kong,* he'd attached a massive inflatable gorilla to the side of the Empire State Building. It had made the front of *The New York Times,* the first time the paper had ever featured an ad on its front page.

"Maybe you've seen the movie *Airplane!* I guess that was before your time, but it's a famous film. With Leslie Nielsen? You know the scene where the pilot and the copilot get food poisoning, and the stewardess switches on the autopilot, and it's an inflatable doll of a pilot? That was me. I made the inflatable autopilot in *Airplane!*"

It was strange, I thought, how his advertising career had brought him into contact with these two classic disaster-based entertainments, before taking him into the further reaches of catastrophe, projected now onto the real world.

"In a way," I said, "you're still in advertising."

"Nah," he said, wise to my angle. "Not at all."

I ventured then that he was perhaps more an insurance salesman, but here, too, he demurred. What he was doing, he said, was getting people somewhere. This place here, he said, and his luxury bunkers in Indiana and Colorado and Germany: they weren't what he was really offering people.

"It's like, you flew here, right? Now tell me this: Was the

airplane the point of your trip? No. You were a little uncomfortable, probably bored. The food sucked. It's a long flight. But you put up with it. Because it got you here. That's what it's like with Vivos. Vivos is the flight. But it's not about the flight. It's about the destination."

Vicino had at his disposal a lavish prospectus of end-time scenarios, an apocalypse to suit every aesthetic taste and ideological preference. Back on the ranchland, as he drove Jin and me across the broken roads into the deeper reaches of the former ordnance depot, he outlined some of these scenarios. There was the "crazy little man in North Korea," and the nuclear war he seemed on the verge of initiating. There was the prospect of hackers, motivated by political aims or pure demonic mischief, unleashing a virus on the systems that controlled the national grid, taking out society's entire technological infrastructure. There were the massive solar flares that occurred periodically and could just as easily do the same thing without need of human agency. He was fond of mentioning the so-called Carrington Event, a massive eruption on the surface of the sun that had occurred at the turn of the last century and caused the breakdown of electrical systems across the world. It wasn't a big deal back then, of course, for the simple reason that we didn't yet have much in the way of electrical systems, but how much worse would it be now, at a time when the wipeout of the grid would mean the near certain collapse of the complex structures undergirding our world?

"And let me tell you, we're well overdue one of those things," he said. "Well overdue."

A major element of Vicino's sales pitch was the idea that the government knew some cataclysmic event was in the offing, but was covering it up to avoid mass panic. You could be sure, he insisted, that those who controlled the world were making arrangements to protect themselves, and that they were hiding from us both those arrangements and the cataclysm itself.

He had some strange beliefs, Vicino, beliefs that were supplementary to his basic apocalyptic vision. He believed that the Earth had a tendency to shift abruptly on its axis, causing massive earthquakes and tidal waves. He believed in the existence of a rogue planet the size of Jupiter called Niburu, which was out there just roaming around untethered to any particular solar system, and that it was on a collision course with our own world, and that the government knew about this, too, and was hiding it from us. He believed that everything that happened, from North Korea to Brexit, was orchestrated with the intention of bringing us closer to one world government.

He wasn't particularly evangelical about these beliefs. He was mostly just putting them out there, it seemed, in the knowledge that apocalyptic unease was basically a volume game. If you didn't like one terrifying dystopian scenario, he had another that might be more your thing. But conspiracies—secret knowledge, hidden revelations—were a key component of his business model. And I was not surprised to glimpse, in due course, the ancient shifting heft of the ur-conspiracy itself. He had, in other words, some strange but familiar notions about the Jews. It was his earnest contention, for instance, that the Democratic Party was essentially a Jewish institution. It had long been the case, he insisted, that a disproportionate number of the party's leaders were Jewish. His historical theory was that because there weren't enough Jews in America to provide them with a constituency sufficient to maintain power, they decided they needed to get the minorities in their corner by promising them handouts.

I sought to clarify that what he was saying was that the Democratic Party was essentially a Jewish conspiracy to gain power through the exploitation of gullible minorities.

"Listen," he said, suddenly wary. "I'm not even slightly anti-Semitic. I banged a lot of Jewish girls in my day. My wife is Jewish, for Christ's sake. My son is half Jewish. So I'm not being anti-Semitic at all when I say that they are very clever, the Jews. Whatever is going on with the breeding they have there. Very clever."

He lifted a great fleshy hand to his graying goatee, massaged it with rhetorical vigor. I took a moment to absorb the physical spectacle of the man. The chunky gold sovereign ring. The beige cargo shorts. The brown leather slip-ons. The pale and weirdly delicate ankles. There was something grimly compelling about all of it. And if my portrayal of him seems to be verging on the mode of caricature, even of outright grotesquerie, it is only because this was how he presented himself to me in fact.

After a further half hour or so of aimless driving, the purpose of which as far as I could see was simply to illustrate the immensity of the property, Vicino brought the Lexus to a stop by another of the vaults. You could see where the prairie winds had blown away some of the topsoil and grass from the top of the structure, revealing an arc of bitumen coating beneath. Ready to get out, I swung open the car door.

The question I needed to ask myself, he said, is which group I wanted to be in when it all went down, when whatever was going to happen happened. When the asteroid landed. When the lights went out. When the economy crashed for good. When the bombs started falling. When the seas began to drown the cities. When the waters were made bitter. When the EMP hit. When for whatever reason, in whatever way, the whole setup went tits irrevocably up, as it unquestionably would. Did I want to be out there trying to get in? Because if I thought I was going to be able to get past the armed guards Vivos would have stationed at all the property's perimeters, good luck to me. I was going to be out there, and you know who was going to be out there with me? A whole lot of other people, and not a lot of food. And it was a known fact, historically, that after twenty-one days without food, people will resort to cannibalism. Here he appealed to the precedent of the Donner Party. In his telling, it seemed to me to be both origin myth and prophecy, a story about a country founded in savagery and destined to devour itself.

"There's going to be gangs roaming," he said. "Cannibals in great numbers. Raping. Pillaging. The have-nots coming after the haves for everything they've got. And my question to you is, do you want your daughters to live through that?"

I did not at that time have any daughters, but I felt it would have been somehow pedantic to point that out, because I understood that on some level he wasn't even talking to me. Like the DIY preppers for whom he professed to have little time, he was speaking of, and to, a conjured phantasm of idealized masculinity—the man who provides, the man who protects, and whom only the breakdown of the state, the collapse of civilization itself, could bring to its truest apotheosis. He was speaking of a man for whom society as a whole had on some level always been a hoard of marauding cannibals baying for the good white Christian flesh of his daughters. The apocalypse, in this sense, was an unveiling of how things really were in this life: of what people were, of what society was, and of how a man stood in relation to it all. Apocalypse, after all, means only this: a revelation, an uncovering of the truth.

It seemed to me that this scenario Vicino had outlined, the haves battening down the hatches against the have-nots, was in some basic sense how the world essentially was, only more so. And though I was not certain about much, I was certain that I didn't want to be one of the haves in a world like that. I knew there was some real hypocrisy in this: if this was already the arrangement of the world, after all, I was nothing if not a have. How could I be so sure that in the wake of some cataclysmic event, I would not be—would not, in fact, have to be—even more heedless of the suffering of others than I already was. Every single day in Dublin I practically stepped over the literal human bodies of the poor, the addicted, the destitute. I complained about the government that did nothing for these people, that had no intention of addressing the systemic injustices that necessitated their suffering, but I myself did essentially nothing to help them, aside from the occasional tossed coin, offered as much to alleviate my own guilt as to ease the suffering of the recipient. But in the end, it was absolutely true that I felt nothing but horror for the product Vicino was trying to sell me, or sell through me. A civilization that could accommodate a business like Vivos was a civilization that had in some sense already collapsed.

During the Cold War years, the fallout bunker was buried deep in the American mind. At a public and private level, in the interlocking languages of politics and pop culture, the prospect of total nuclear annihilation embedded itself in the discourse, as an ever-present possibility, even an outright likelihood. A few weeks after the destruction of Hiroshima, Bertrand Russell spoke of the probability, in the coming years, of the absolute obliteration of humanity and its works. "One must expect a war between U.S.A. and U.S.S.R.," he wrote, "which will begin with the total destruction of London. I think the war will last 30 years, and leave a world without civilised people, from which everything will have to be built afresh—a process taking (say) 500 years." Twenty-four years later, in an interview with *Playboy,* he had not encountered much in the way of cause for increased optimism: "I still feel that the human race may well become extinct before the end of the present century. Speaking as a mathematician, I should say that the odds are about three to one against survival."

In his book *One Nation Underground: The Fallout Shelter in American Culture,* Kenneth D. Rose writes about the Cold War–era proliferation of detailed scenarios of nuclear annihilation in major newspapers and magazines:

> Like other genres, the nuclear apocalyptic, whether found in a specialty or mainstream publication, had

> certain conventions that were observed by its practitioners [. . .] Descriptions of radiation sickness, blindness, terrible burns, gaping wounds, and missing limbs proliferate, and the authors conjure up gruesome images of corpses strewn about homes, workplaces, and sidewalks. When the bombs finally stop falling, the nuclear apocalyptic will emphasize how life has been reduced to a low, primitive state in which each day is a grim struggle for survival. Disease, dementia, lawlessness, and hunger are constant companions. In many nuclear apocalyptics, an occasional artifact from the past will surface to remind survivors of all that they have lost. Finally, depending on the author's view of nuclear war, the survivors will either begin building a new tomorrow, rising phoenix-like from the ashes, or hopelessly resign themselves to an endless era of darkness and barbarity.

In July 1961, with Khrushchev threatening to negotiate a new peace deal with East Germany, thereby declaring Berlin a "neutral" city and forcing the US Army to withdraw, John F. Kennedy delivered a speech, broadcast on national television, that redoubled the American public's fears of nuclear war. "We do not want to fight," he said, "but we have fought before." In the event of a Soviet attack, "the lives of those families which are not hit in a nuclear blast and fire can still be saved—if they can be warned to take shelter, and if that shelter is available. . . . The time to start is now," he said. "In the coming months, I hope to let every citizen know what steps he can take without delay to protect his family in case of attack. I know you would not want to do less." After the speech, Congress voted to allocate $169 million for the construction and stocking of fallout shelters in public and private properties across the country.

The number of families who went on to build shelters was relatively low, but the trend was seen, among cultural critics of the time, as indicative of the fearful self-enclosure of the new suburban middle class. Writing in *The New York Times* in 1961, the anthropologist Margaret Mead suggested that anxieties about nuclear war and urban crime had caused a mass retreat to the suburbs, where middle-class Americans were hiding from the world and its darkening future. "The armed, individual shelter," she wrote, "is the logical end of this retreat from trust in and responsibility for others."

I have some sympathy for the builders of bunkers, the hoarders of freeze-dried foodstuffs. I understand the fear, the desire for it to be assuaged. But more than I want my fear assuaged, I want to resist the urge to climb into a hole, to withdraw from an ailing world, to bolt the door after myself and my family. When I think of Vicino's project, his product, what comes to mind is exactly Margaret Mead's judgment of what it means to secure oneself inside a shelter: a withdrawal from any notion that our fate might be communal, that we might live together rather than survive alone.

The bunker, purchased and tricked out by the individual consumer, is a nightmare inversion of the American Dream. It's a subterranean abundance of luxury goods and creature comforts, a little kingdom of reinforced concrete and steel, safeguarding the survival of the individual and his family amid the disintegration of the world.

"Burying your head in the sand," Vicino told me, "isn't gonna save your ass that's hanging out." He was, he said, paraphrasing Ayn Rand—his point being, I supposed, that not purchasing a place in one of his facilities amounted to an unwillingness to face down the reality of the world. But it struck me as an odd image, an odd analogy to use, given what he himself was advocating for. Because as far as I understood him, what he was arguing was this: it was no use burying your head in the sand if you didn't also bury your ass along with it.

The following day I returned to xPoint. Neither Vicino nor Jin was in evidence, and there was no sign of the Lexus. Outside the corrugated iron shipping container was an unmanned red jeep with decals on its doors advertising a local Fox News station, and I inferred that Vicino had gone tooling around the prairies with a TV reporter, perhaps adjusting his sales pitch to the particular anxieties of a South Dakota conservative cable news–viewing demographic. I parked beside the jeep and set out to wander the site, but then quickly realized it was far too vast to even begin to explore on foot and returned to the car. I drove for forty minutes or so, stopping now and then to unlock a cattle gate, and once or twice to get out and observe the delirious spectacle of the endless grass-covered vaults, the hexagonal fronts—an architecture less proportionate to the physical than to the psychic dimensions of human beings. I clambered up onto the top of one of these, to survey the immensity from a higher vantage. Yesterday, Jin and I had stood on top of another of these structures, and my growing apprehension of a military-industrial sublime had been casually undermined by Jin's solemnly informing me that he'd recently taken a shit on the roof of one such vault, although "probably not this one."

I sat down now on the sparse grass of the roof and looked out across the infinity of green, surreally ruptured by the vaults. The thought occurred to me that it was here, in what was then the southern part of the Dakota Territory, that Laura Ingalls Wilder spent much of her childhood, and where she set several of her *Little House* novels. This was not just a prairie I was looking out over, therefore, but *the* prairie: the fertile source of America's dream of itself as a nation of entrepreneurial pioneers, settlers of a wild land. I was looking out over a country born in savagery and genocide, built on the ruins of a conquered native civilization, and the bunkers seemed to me like the return of a repressed apocalypse. It was as though the land itself had extruded them as an immune response to some ancient antigen. It was so quiet here I could hear the soft buzzing of electricity in the power lines above me, the brittle snap and hum of technological civilization itself. I thought about America's twin obsessions with a frontier past and an apocalyptic future, and of how these were ominously fused like the Janus-faced calf at the Pioneer Museum. What was Vicino offering in this place, after all, other than a return to the life of the old frontier, a new beginning in the wake of the end, one that retained as many consumer-facing luxuries as possible?

I saw a plume of dust drifting on the horizon, heralding the approach of Vicino's Lexus. By the time I reached the shipping container, a young woman in a yellow cocktail dress was holding a microphone and delivering a monologue to a camera she'd set up beside an empty barn.

"If and when the end comes," she intoned, "will you be ready?"

I sidled up to Vicino, who nodded at me and asked by way of openers what number out of ten I'd assign to the reporter in terms of sexual desirability. I told him I wouldn't really want to say, and he shrugged and said that he personally felt she'd suffice in an apocalyptic lockdown scenario, the implication being that she would by no means otherwise be to his taste.

I wondered whether this was a deliberate provocation on his part, one of the running jokes he had going with himself, and imagined him bragging later about how he'd outraged the delicate sensibilities of this lefty European writer by continually asking him to judge the physical attractiveness of various women on a numerical rating scale.

Driving east across the prairies, I wondered what it might mean to think of Vicino as, if not a savior as such, then a man who happened to be in a position to offer salvation. The idea used to be that God would spare the righteous while the ungodly perished. These matters now were in the hands of the market. If you could afford the outlay, and if you had the foresight to get in on the ground floor, you were in with a chance to be among the saved. That was business: the first and the last, the alpha and omega.

When I got to Hot Springs it was evening, and the sun was flooding the western sky with a golden light. I drove around for a while, trying to find somewhere to eat that wasn't Pizza

Hut. Main Street was entirely deserted but for the solitary and oblivious figure of a young man, standing in an empty parking space by the river, across the street from an elegantly dilapidated old movie house. He had on a backward baseball cap and a black vest, and he had lank blond hair down to his shoulders. He was extremely pale and a little pudgy, and his eyes were closed, and his legs were spread in a power stance of heavy-metal defiance. He was shredding an intangible guitar, working his fingers up and down the frets in a blur of agile certitude, thrashing his head in righteous assent with whatever music was playing on his headphones.

I slowed down as I passed him, staring openly through the side window at this spectacle of pure face-melting abandon. There was something both frightening and life affirming about the performance, which was hardly a performance at all, in that it was presumably delivered entirely for the performer's own personal gratification. I drove on and pulled into a parking spot a little way up the street, and sat there awhile, trying to make sense of what I'd just seen, to fit it into some framework of meaning or significance. I walked back up toward him then, on the opposite side of the street, and when he came back into view I could see that he had abandoned the guitar and had moved on to vocals.

I leaned against the window of a closed military surplus store and took stock of him for a while. He would stand bent as if in prayer over an invisible microphone, mouthing along silently and with shocking intensity to an unheard music, and then abandon himself to an improvised dance, a furious rhythmical stamping that made me think of Rumpelstiltskin.

There was real violence to it, real chaos and fury. A car rounded the corner and passed him by, and neither of its occupants, a man and woman in middle age, seemed to pay him any heed at all. Perhaps they knew him, I thought, and were familiar with these desperate displays of pure energy, of wild and illegible life. And if so, I thought, they had surely long relinquished the need for it to mean or symbolize anything.

# 4

# BOLT-HOLE

Within about an hour of arriving in Auckland, I was as close to catatonic from fatigue as made no difference, and staring into the maw of a volcano. I was standing next to an art critic named Anthony Byrt. He'd picked me up from the airport and, in a gesture I would come to understand as quintessentially Kiwi, dragged me directly up the side of a volcano. This particular volcano, Mount Eden, was a fairly domesticated specimen, around which was spread one of the more affluent suburbs of Auckland.

I was a little out of breath from the climb and, having just emerged in the southern hemisphere from a Dublin November, sweating liberally in the relative heat of the early summer morning. I was also experiencing near-psychotropic levels of jet lag. I must have looked a bit off, because Anthony offered a cheerful apology for playing the volcano card so early in the proceedings.

"I probably should have eased you into it, mate," he chuckled. "But I thought it'd be good to get a view of the city before breakfast."

Anthony was in his late thirties, neatly bearded with a shaven head. His manner was fluently loquacious and at the same time politely diffident.

The view of Auckland and its surrounding islands was indeed ravishing—though, in retrospect, it was no more ravishing than any of the countless other views I would wind up getting ravished by over the next ten days. That, famously, is the whole point of New Zealand: if you don't like getting ravished by views, you have no business in the place. To travel there is to give implicit consent to being hustled left, right and center into states of aesthetic rapture.

"I've been in the country mere minutes," I said, "and I've already got a perfect visual metaphor for the fragility of civilization in the bag." I was referring here to the pleasingly surreal spectacle of a volcanic crater overlaid with a surface of neatly manicured grass.

New Zealand, Anthony pointed out, was positioned right on the Pacific Ring of Fire, a horseshoe curve of geological fault lines stretching upward from the western flank of the Americas, back down along the eastern coasts of Russia and Japan and on into the South Pacific. It was volcano country.

It seemed odd, I said, given all this seismic activity, that superrich Silicon Valley technologists were supposedly apocalypse-proofing themselves by buying up land down here.

"Yeah," said Anthony, "but some of them are buying farms and sheep stations pretty far inland. Tsunamis aren't going to be a big issue there. And what they're after is space, and clean water. Two things we've got a lot of down here."

It was precisely this phenomenon—of tech billionaires buying up property in New Zealand in anticipation of civilizational collapse—that constituted our shared apocalyptic obsession. This was the reason I had come down here, to find out about these apocalyptic retreats, and to see what New Zealanders thought of this perception of their country.

In any discussion of our anxious historical moment, its apprehensions of decay and collapse, New Zealand was never very far from being invoked. It was the ark of nation-states, an island haven amid a rising tide of apocalyptic unease: a wealthy, politically stable country unlikely to be seriously affected by climate change, a place of lavish natural beauty with vast stretches of unpopulated land, clean air, fresh water. For those who could afford it, New Zealand offered at the level of an entire country the sort of reassurance promised by Vicino's bunkers.

According to New Zealand's Department of Internal Affairs, in the two days following the 2016 election, the number of Americans who visited its website to inquire about the process of gaining citizenship increased by a factor of fourteen compared with the same day in the previous month. That same week, *The New Yorker* ran a piece about the superrich making preparations for a grand civilizational crack-up. Speaking of New Zealand as a "favored refuge in the event of a cataclysm," Reid Hoffman, the founder of LinkedIn, claimed that "saying you're 'buying a house in New Zealand' is kind of a wink, wink, say no more."

And then there was Peter Thiel, the billionaire venture capitalist who had cofounded PayPal and had been one of Facebook's earliest investors. It had recently emerged that

Thiel had bought a sprawling property in New Zealand, on the shores of Lake Wanaka, the apparent intention of which was to provide him with a place to retreat to should America become unlivable due to economic chaos, civil unrest, or some or other apocalyptic event. (Sam Altman, one of Silicon Valley's most influential entrepreneurs, alluded in an interview to an arrangement with his friend Thiel, whereby in the eventuality of some kind of systemic collapse scenario—synthetic virus breakout, rampaging AI, resource war between nuclear-armed states, and so forth—they both get on a private jet and fly to this property in New Zealand. The plan from this point, you'd have to assume, was to sit out the collapse of civilization before re-emerging to provide seed funding for, say, the insect-based protein sludge market.)

In the immediate wake of Altman's revelation about the New Zealand apocalypse contingency plan, a reporter for the *New Zealand Herald* named Matt Nippert had begun looking into the question of how exactly Thiel had come into possession of this 477-acre former sheep station in the South Island, the larger and more sparsely populated of the country's two major landmasses. Foreigners looking to purchase significant amounts of land in New Zealand typically had to pass through a stringent government vetting process, but in Thiel's case, Nippert learned, no such process had been necessary. The reason for this, he revealed, was that Thiel was in fact a citizen of New Zealand, despite having spent no more than twelve days in the country up to that point and having not been seen in the place since 2011. He hadn't even needed to make the trip to New Zealand to have his citizenship confirmed, it turned out:

the deal had been sealed in a private ceremony at a consulate handily located, for Thiel's purposes, in Santa Monica.

Everyone was always saying these days that it was easier to imagine the end of the world than the end of capitalism. Everyone was always saying it, in my view, because it was obviously true. The perception, paranoid or otherwise, that billionaires were preparing for a coming collapse seemed a literal manifestation of this axiom. Those who were saved, in the end, would be those who could afford the premium of salvation. And New Zealand, in this story, had become a kind of contemporary Ararat: a place of shelter from the coming flood.

I myself could not help taking all this personally. Reading about these billionaires and their plans to protect themselves and their money while the rest of us burned, I felt an almost visceral revulsion for these people, and for the system that afforded them such disproportionate wealth and power. Like Vicino's bunkers, this seemed to me to represent a radical acceleration of the mechanisms by which our civilization was already driven.

Thiel, in this sense, loomed particularly large. Through his data analytics company Palantir, he was a presiding presence in the increasingly oppressive, though only fleetingly visible, environment of surveillance capitalism. He was known for his extreme libertarian views. "I no longer believe," he had once written, "that freedom and democracy are compatible." His conception of freedom had less to do with existential liberty, with meaningful human lives within flourishing communities, than it had to do with not having to share resources—the freedom of wealthy people from taxation, from any obligation to materially contribute to society. And he was known, too, for his apparent determination to literally live forever, through investing in various life extension therapies and technologies. (As though enthusiastically pursuing the clunkiest possible metaphor for capitalism at its most vampiric, he had publicly expressed interest in a therapy involving regular transfusions of blood from young people as a potential means of reversing the aging process.)

He was in one sense a figure of almost cartoonishly outsized villainy. But in another, deeper sense, he was pure symbol: less an actual person than a shell company for a diversified portfolio of anxieties about the future, a human emblem of the moral vortex at the center of the market. It was in this second sense that I was fascinated and horrified by Thiel, who seemed to me increasingly to represent the world my son would likely be forced to live in.

It was in the early summer of 2017, just as my interests in the topics of New Zealand and Thiel and civilizational collapse were beginning to converge into a single obsession, that I first heard from Anthony. He had read a book of mine that had been published earlier that year, an account of the transhumanists of Silicon Valley and their obsession with achieving immortality through technological means, and had recognized in my writing about Thiel something of his own personal fascination with the man.

We began a long and intricate exchange of emails, largely

focused around Thiel and his attraction to New Zealand. If I wanted to understand the extreme ideology that underpinned this attraction, he told me, I needed to understand an obscure libertarian manifesto called *The Sovereign Individual: How to Survive and Thrive During the Collapse of the Welfare State.* It was published in 1997, and in recent years something of a minor cult had grown up around it in the tech world, largely as a result of Thiel's citing it as the book he was most influenced by. Other prominent boosters included Netscape founder and venture capitalist Marc Andreessen and Balaji Srinivasan, the entrepreneur best known for advocating Silicon Valley's complete secession from the United States to form its own corporate city-state. *The Sovereign Individual*'s coauthors were James Dale Davidson, a private investor who specializes in advising the rich on how to profit from economic catastrophe, and the late William Rees-Mogg, long-serving editor of the *Times* and father of Jacob Rees-Mogg, the Conservative MP beloved of Britain's reactionary pro-Brexit right.

I was intrigued by Anthony's description of the book as a master key to the relationship between New Zealand and the techno-libertarians of Silicon Valley. Reluctant to enrich Davidson or the Rees-Mogg estate any further, I bought a used edition online, the musty pages of which were here and there smeared with the desiccated snot of whatever nose-picking libertarian had preceded me. It presented a bleak vista of a post-democratic future. Amid a thicket of analogies to the medieval collapse of feudal power structures, the book also managed, a decade before the invention of Bitcoin, to make some impressively accurate predictions about the advent of online economies and cryptocurrencies. Its four-hundred-odd pages of near-hysterical orotundity can roughly be broken down into the following sequence of propositions:

1) The democratic nation-state basically operates like a criminal cartel, forcing honest citizens to surrender large portions of their wealth to pay for stuff like roads and hospitals and schools.
2) The rise of the Internet, and the advent of cryptocurrencies, will make it impossible for governments to intervene in private transactions and to tax incomes, thereby liberating individuals from the political protection racket of democracy.
3) The state will consequently become obsolete as a political entity.
4) Out of this wreckage will emerge a new global dispensation, in which a "cognitive elite" will rise to power and influence, as a class of sovereign individuals "commanding vastly greater resources" who will no longer be subject to the power of nation-states and will redesign governments to suit their ends.

*The Sovereign Individual* is, in the most literal of senses, an apocalyptic text. Davidson and Rees-Mogg present an explicitly millenarian vision of the near future: the collapse of old orders, the rising of a new world. Liberal democracies will die out and be replaced by loose confederations of corporate city-states. Western civilization in its current form, they insist, will end with the millennium. "The new Sovereign

Individual," they write, "will operate like the gods of myth in the same physical environment as the ordinary, subject citizen, but in a separate realm politically." It's impossible to overstate the darkness and extremity of the book's projected future. To read it is to be continually reminded that the dystopia of your darkest insomniac imaginings is almost always someone else's dream of a new utopian dawn.

Davidson and Rees-Mogg identified New Zealand as an ideal location for this new class of sovereign individuals, as a "domicile of choice for wealth creation in the Information Age." Anthony, who drew my attention to these passages, had even turned up evidence of a property deal in the mid-1990s in which a giant sheep station at the southern tip of the North Island was purchased by a conglomerate whose major shareholders included Davidson and Rees-Mogg. Also in on the deal was one Roger Douglas, New Zealand's former finance minister, who had presided over a radical restructuring of the country's economy along neoliberal lines in the 1980s. (This period of so-called "Rogernomics," Anthony told me—the selling off of state assets, slashing of welfare, deregulation of financial markets—created the political conditions that had made the country such an attractive prospect for wealthy Americans.)

Thiel was famously obsessed with the work of J. R. R. Tolkien, and his interest in New Zealand was not unconnected with the fact that Peter Jackson's film adaptations of *The Lord of the Rings* had been filmed there. This was a man who had named at least five of his companies in reference to Middle Earth and fantasized as a teenager about playing chess against a robot that could discuss the books. It was a matter, too, of the country's abundance of clean water and the convenience of overnight flights from California. But it was also inseparable from a particular strand of apocalyptic libertarianism. To read *The Sovereign Individual* was to see this ideology laid bare: these people, the self-appointed "cognitive elite," were content to see the unraveling of the world as long as they could carry on creating wealth in the end times.

I was struck by how strange and disquieting it must have been for a New Zealander to see their own country refracted through this strange apocalyptic lens. It was Anthony's contention that if I was interested in the end of the world, I had to understand the relationship between his country and the Silicon Valley tech elite. He was also of the opinion that if I wanted to understand it, I had to go there. And so somewhere in the course of our email exchange, a plan started to formulate. I was going to travel to New Zealand, and we were going to take a trip to Peter Thiel's apocalypse retreat on the shores of Lake Wanaka.

After the hike up to Mount Eden, Anthony dropped me off at my hotel. I dumped my bags and I wandered the streets of downtown Auckland awhile. My jet lag had by then progressed into a kind of fugue state. I was operating on existential factory settings, reduced to the barest human functionality. Realizing that I hadn't eaten in something like twelve hours, and that I was in fact ravenously hungry, I found myself drifting, as though without awareness or volition, toward the familiar sight of a Nando's. I placed my order and sat down, and was immediately struck by the absurdity of my having flown twenty-six straight hours to this archipelago in the far southwestern Pacific, only to find myself sitting in a Nando's that was in no way distinguishable from the Nando's ten minutes' walk from my house in Dublin, a restaurant I would by no means have walked ten minutes to eat at.

I was waiting for my flame-grilled chicken thighs and spicy rice to materialize, groping the while toward some vague insight about how globalization was continuing the old colonial pursuit of flattening and assimilation, when I became aware of a small bird perched on the back of the seat opposite me. At first I thought I might have been imagining it, that the jet-lag fugue state had now progressed to outright visual hallucination, but then this small bird, perhaps a thrush or a sparrow, took to the air and in the course of its flight caused a young woman at a nearby table to duck slightly, which seemed evidence enough that my senses did not deceive me. Presently, another bird flew in off the street and took its ease briefly on a napkin dispenser at an empty table before taking flight and circling the restaurant in the wake of its companion. Nobody seemed to be paying the least bit of attention to these birds treating the inside of a Nando's like some kind of aviary. It occurred to me then that New Zealand was the last country in the world to be settled by humans, that until the Māori arrived here in the thirteenth century no mammal of any sort had ever existed on this land, and that in the absence of large predators the whole place had been until then essentially a giant bird sanctuary.

This was something I noticed on numerous occasions in Auckland. I would be sitting in a restaurant or café, and there would be birds just flying around, alighting more or less unobtrusively on the backs of people's chairs, pecking at crumbs beneath their tables, and no one else would seem to notice. It was strange, but also strangely wonderful. One evening, over dinner with Anthony and his wife, Kyra, at their house, I mentioned this business of birds just flying into restaurants and cafés like they owned the place, and they both seemed at a loss, as though they had not themselves paid it much notice, or were unaware that it was a uniquely Kiwi phenomenon, and I found this touching in a way I couldn't quite articulate. It had to do, I think, with a sense of New Zealand not just as a place of extraordinary natural beauty, but also as retaining some quality of original innocence (a perception of the place that, when I reflected on it, seemed to me tainted by a colonialist view of the world). It hadn't in the scheme of things been that long since the birds had indeed owned the place, and it was as though they still hadn't quite adjusted to the new arrangement. I was beginning already to see how New Zealand could create in the newly arrived traveler a sense of having arrived in a place that was both recognizably of the Anglophone West and at the same time somehow prelapsarian. A place where you could eat at Nando's while small birds alighted on your table, harmless and unafraid.

The following day, I went to a gallery in downtown Auckland to take a look at a new work by the artist Simon Denny.

Anthony and I had talked a lot about this exhibition, because it touched on many of our shared fascinations, and because it was a project he himself had been involved in from its inception. (He had written some enthusiastic criticism on Denny's work, and the two had started a correspondence, which eventually opened out onto the prospect of some kind of collaboration. Anthony characterized his own role in the project as an amalgamation of researcher, journalist, and "investigative philosopher, following the trail of ideas and ideologies.") The exhibition was called *The Founder's Paradox*, a name that came from the title of one of the chapters of Peter Thiel's 2014 book *Zero to One: Notes on Startups, or How to Build the Future.* Together with the long and intricately detailed catalog essay Anthony was writing to accompany it, the show was a reckoning with the future that Silicon Valley techno-libertarians like Thiel wanted to build, and with New Zealand's place in that future. *The Sovereign Individual*, too, was a central element of the show.

When I got to the gallery, Simon Denny, whom Anthony had described to me as "kind of a genius" and "the poster-boy for post-Internet art," was making some last-minute preparations for the show's opening. He was a neat and droll man in his mid-thirties, a native of Auckland who had lived for many years in Berlin, where he was a significant figure in the international art scene. He talked me through the conceptual framework for the show. It was structured around games—in theory playable, but in practice encountered as sculptures—representing two different kinds of political vision for New Zealand's future. The bright and airy ground floor space was filled with tactile, bodily game-sculptures, riffs on Jenga and

Operation and Twister. These works, incorporating collaborative and spontaneous ideas of play, were informed by a recent book called *The New Zealand Project* by a young left-wing thinker named Max Harris, which explored a humane, collectivist politics influenced by Māori beliefs about society.

Down in the low-ceilinged, dungeon-like basement was a set of sculptures rooted in an entirely different understanding of play, more rule-bound and cerebral. These were based on the kind of role-playing strategy games particularly beloved of Silicon Valley tech types, and representing a Thielian vision of the country's future. The psychological effect of this spatial dimension of the show was immediate: upstairs, you could breathe, you could see things clearly, whereas to walk downstairs was to feel oppressed by low ceilings, by an absence of natural light, by the apocalyptic darkness captured in Simon's elaborate sculptures.

This was a world Simon himself knew intimately. What was strangest and most unnerving about his art was the sense that he was allowing us to see this world not from the outside in, but from the inside out, and this required a certain level of proximity—often to people whose politics he found repellent. (There was in this sense a journalistic quality in Simon's approach to his work, if not to the work itself.) Over beers in Anthony's kitchen the previous night, Simon had told me about a dinner party he had been to in San Francisco earlier that year, at the home of a techie acquaintance. There had been a lot of Silicon Valley new money types there, he said, a lot of "blockchain entrepreneurs." There were MAGA hats, and there was palpable excitement about Trump and the great rupture he seemed to represent. These people were from hacker backgrounds, and their view of the world arose out of a deep ethos of lulz. It was as though the new president had pulled off the ultimate troll on the liberal establishment. Seated next to Simon at dinner was a man named Curtis Yarvin, who had founded a computing platform named Urbit, with the help of Thiel's money. As anyone who took an unhealthy interest in the weirder recesses of the online far-right was aware, Yarvin was more widely known as the blogger Mencius Moldbug. Moldbug was the intellectual progenitor of neoreaction, an antidemocratic movement that advocated for a kind of white-nationalist oligarchic neofeudalism—rule by and for a self-proclaimed cognitive elite—and which had found a small but influential constituency in Silicon Valley.

Beneath all the intricacy and detail of its world-building, *The Founder's Paradox* was clearly animated by an uneasy fascination with the utopian future imagined by the techno-libertarians of Silicon Valley. The exhibition's centerpiece was a tabletop strategy game called *Founders,* which drew heavily on the aesthetic—as well as the explicitly colonialist language and objectives—of The Settlers of Catan, a massively popular multiplayer strategy board game. The aim of *Founders,* as clarified by the accompanying text and by the piece's lurid illustrations, was not simply to evade the apocalypse, but to prosper from it. First you acquired land in New Zealand, with its rich resources and clean air, away from the chaos and ecological devastation gripping the rest of the world. Next you moved on to seasteading, the libertarian ideal of constructing man-made islands in international waters. On these floating utopian micro-states, wealthy tech innovators would be free to go about their business without interference from democratic governments. (Thiel was an early investor in, and advocate of, the seasteading movement, though his interest has waned in recent years.) Then you mined the moon for its ore and other resources, before moving on to colonize Mars. This last level of the game reflected the current preferred futurist fantasy, most famously advanced by Thiel's former PayPal colleague Elon Musk, with his dream of fleeing a dying planet Earth for privately owned colonies on Mars.

The influence of *The Sovereign Individual* was all over the show. It was a detailed mapping of a possible future, in all its highly sophisticated barbarism. It was a utopian dream that appeared, in all its garish detail and specificity, as the nightmare vision of a world to come. Standing alone in the central chamber of the basement, peering through the glass case at the board of the *Founders* tabletop game, inspecting each of the various illustrated spaces, I noted a familiar image, a hexagonal concrete structure emerging out of grass. It was one of the bunkers at xPoint in South Dakota. I felt as though I were looking at the manifestation of my own anxieties about the future, anxieties that had often seemed to me bewilderingly complex and idiosyncratic, and irretrievably entangled with a revulsion at the cruelty and destructiveness of capitalism. The game represented an apocalyptic logic of progress: a movement away from the nation-state, away from democracy, and finally away from the ravaged Earth itself. It represented everything I thought about when I thought about the end of the world. It was like being confronted with a lurid diorama of my own unease as I had come to conceive of it. It was uncanny, and terrible, and strangely perfect.

Later that week, in a bar a few blocks from the harbor, I had a post-work beer with Matt Nippert, the *New Zealand Herald* reporter who had broken the citizenship story earlier that year. He told me of his personal certainty that Thiel had bought his property in the South Island for apocalypse-contingency purposes. In his citizenship application, he had pledged his commitment to devote "a significant amount of time and resources to the people and businesses of New Zealand." But none of this had amounted to much, Nippert said, and he was convinced that it had only ever been a feint to get him in the door.

I was not surprised to find that the handful of people I spoke to from the luxury property business did not see it this way. They were keen to portray New Zealand as a kind of utopian sanctuary, but to give as little oxygen as possible to the related narrative around the country as an apocalyptic bolt-hole for the international elite. Over coffee at his golf club, Terry Spice—a London-born luxury property specialist who had recently sold a large estate abutting the Thiel property on Lake Wanaka—told Anthony and me that Thiel had highlighted internationally the country's reputation as "a safe haven, and a legacy asset." He himself had sold land to one very wealthy American client who had called him on the night of the presidential election.

"This guy couldn't believe what was happening," he said. "He wanted to secure something right away."

But on the whole, he insisted, this kind of apocalyptically motivated buyer accounted for a vanishingly small proportion of the market.

Showing me around the high-end beachfront properties he represented about an hour or so north of Auckland, another luxury property specialist named Jim Rohrstaff, a Californian transplant who specialized in selling to the international market, likewise told me that although quite a few of his major clients were Silicon Valley types—I wanted him to name names, but he politely responded that he didn't "kiss and tell"—the end of the world tended not to be a particular factor in their purchasing decisions.

"Look," he said, "it might be one strand in terms of what's motivating them to buy here. But in my experience it's never been the overriding reason. It's much more of a positive thing. What they see when they come here is utopia."

Thiel himself had spoken publicly of New Zealand as a "utopia," during the period in 2011 when he was maneuvering for citizenship, investing in various local startups under a venture capital fund called Valar Ventures. (Valar, needless to say, was another Tolkien reference.) This was a man with a particular understanding of what a utopia might look like—who did not believe, after all, in the compatibility of freedom and democracy. In a *Vanity Fair* article about his role as adviser to Donald Trump's presidential campaign, a friend was quoted as saying that "Thiel has said to me directly and repeatedly that he wanted to have his own country," adding that he had even gone so far as to price up the prospect at somewhere around one hundred billion dollars.

The Kiwis I spoke with were uncomfortably aware of what Thiel's interest in their country represented, of how it seemed to figure more generally in the frontier fantasies of American libertarians. Max Harris, the author of *The New Zealand Project,* the book that had informed the game sculptures on the upper level of *The Founder's Paradox,* mentioned that for much of its history the country tended to be viewed as a kind of political Petri dish—it was, for instance, the first nation to recognize women's right to vote—and that this "perhaps makes Silicon Valley types think it's a kind of blank canvas to splash ideas on."

When we met in her office at the Auckland University of Technology, the legal scholar Khylee Quince insisted that any invocation of New Zealand as a utopia was a "giant red flag," particularly to Māori like herself. "That is the language of emptiness and isolation that was always used about New Zealand during colonial times," she said. And it was always, she stressed, a narrative that erased the presence of those who were already here: her own ancestors.

The first major colonial encounter for Māori in the nineteenth century was not with representatives of the British crown, she pointed out, but with private enterprise. The New Zealand Company was a private firm founded by a convicted English child kidnapper named Edward Gibbon Wakefield, with the aim of attracting wealthy investors with an abundant supply of inexpensive labor—migrant workers who could not themselves afford to buy land in the new colony, but who would travel there in the hope of eventually saving enough wages to buy in. The company embarked on a series of expeditions in the 1820s and '30s. It was only when the firm started drawing up plans to formally colonize New Zealand, and to set up a government of its own devising, that the British colonial office advised the crown to take steps to establish a formal colony. In the utopian fantasies of techno-libertarians like Thiel, Quince saw an echo of that period of her country's history.

"Business," she said, "got here first."

Given her Māori heritage, Quince was particularly attuned to the colonial resonances of the more recent language around New Zealand as both an apocalyptic retreat and a utopian space for American wealth and ingenuity.

"I find it incredibly offensive," she said. "Thiel got citizenship after spending twelve days in this country, and I don't know if he's even aware that Māori exist. We as indigenous people have a very strong sense of intergenerational identity and collectivity. Whereas these people, who are sort of the contemporary iteration of the colonizer, are coming from an ideology of rampant individualism, rampant capitalism."

On New Zealand's left, there had lately been a kindling of cautious optimism, sparked by the recent surprise election of a new Labour-led coalition government, under the leadership of the thirty-seven-year-old Jacinda Ardern, whose youth and apparent idealism seemed to suggest a move away from neoliberal orthodoxy. During the election, foreign ownership of land had been a major talking point, though it focused less on the wealthy apocalypse-preppers of Silicon Valley than on the perception that overseas property speculators were driving up the cost of houses in Auckland. The incoming government had committed to tightening regulations around land purchases by foreign investors, and would eventually make it much more difficult for foreign buyers to gain a foothold in the property market. This was largely the doing of Winston Peters, a populist of Māori descent whose New Zealand First party held the balance of power and who was strongly in favor of tightening regulations of foreign ownership. Peters had been a prominent figure in Kiwi politics since the 1970s. When I read that Ardern had named Peters as her deputy prime minister, I was surprised to recognize the name—from, of all places, *The Sovereign Individual,* where Davidson and Rees-Mogg had singled him out for weirdly personal abuse as an archenemy of the rising cognitive elite, referring to him as a "reactionary loser" and "demagogue" who would "gladly thwart the prospects for long-term prosperity just to prevent individuals from declaring their independence of politics."

During my time in New Zealand, Ardern was everywhere: in the papers, on television, in every other conversation. On our way to Queenstown in the South Island, to see for ourselves the site of Thiel's apocalyptic retreat, Anthony and I were in the security line at Auckland Airport when a woman of about our age, smartly dressed and accompanied by a cluster of serious-looking men, glanced in our direction as she was conveyed quickly along the express lane. She was talking on her phone but looked toward us and waved at Anthony, smiling broadly in happy recognition.

"Who was that?" I asked.

"Jacinda," he said.

"You know her?"

"We know quite a lot of the same people. We met for a drink a couple of times back when she was Labour's arts spokesperson."

"Really?"

"Well, yeah," he laughed, "there's only so many of us."

"The endgame for Thiel is essentially *The Sovereign Individual,*" said Anthony. He was driving the rental car, allowing me to fully devote my resources to the ongoing cultivation of aesthetic rapture (mountains, lakes, so forth). We were on our way to see for ourselves the part of New Zealand, on the shore of Lake Wanaka in the South Island, that he had bought for purposes of post-collapse survival. We talked about the trip as though it were a gesture of protest, but it felt like a kind of perverse pilgrimage. The term *psychogeography* was cautiously invoked, and with only the lightest of ironic inflections.

"And the bottom line for me," he said, "is that I don't want my son to grow up in that future."

I had met Anthony's son, a clever and charming and prodigiously chatty seven-year-old, and I didn't want him to grow up in that future either. This was one of the things we had bonded over, Anthony and I, the fact that we were both fathers of young boys, had similar concerns about their future. It was supposed to be an inevitable part of life that you drifted gradually but perceptibly rightward when you had kids. You got older, you began to think of yourself as a "centrist." You took up golf, maybe started laying down bottles of wine. But both of us had been radicalized by parenthood. Having children had brought into horribly lurid focus the predatory face of contemporary capitalism.

Symbolically speaking, this face was Thiel's.

"The thing about Thiel is he's the monster at the heart of the labyrinth," said Anthony.

"He's the white whale," I suggested, getting into the literary spirit of the enterprise.

We were joking, but also not. Our shared fixation occupied a kind of Melvillean register, yearned toward a mythic scale. For Anthony, it colored his perception of everything, including his immediate environment. He admitted to a strange aesthetic pathology whereby he encountered, in the alpine grandeur of the South Island, not the sublime beauty of his own home country, but rather what he imagined Thiel seeing in the place: Middle-earth. Thiel's Tolkien fixation was itself a fixation for Anthony: together with the extreme libertarianism of *The Sovereign Individual,* he was convinced that it lay beneath Thiel's continued interest in New Zealand. It was his view that, on some level, the place in which he planned to spend his radically extended postapocalyptic life span was not New Zealand at all, but Middle-earth.

The effect of Peter Jackson's films on the country was strangely all-consuming. The previous evening in Anthony's kitchen in Auckland, we'd been looking up locations we wanted to hit in the South Island, the routes we'd take between them, when we'd discovered that Google Maps allowed you to search for fictional Middle-earth locations—Isengard, Mordor, Hobbiton, the Dead Marshes, Fangorn Forest, and so on—thereby providing you with the outline of a real place on top of which an entirely fictional region had been mapped. It was, in this sense, an uncanny revisiting of the original sin of the colonial encounter. I thought of Borges's story "Tlön, Uqbar, Orbis Tertius," in which the discovery of a hoax encyclopedia from an invented distant world causes the "real" world to give way beneath the pressure of fiction. (It struck me then that the company owned by Curtis Yarvin—the neoreactionary extremist whose software platform Thiel had funded, and whom Simon had told me about sitting beside at dinner in San Francisco earlier that year—was called Tlön, and that its stated aim was "to build a new internet on top of the old internet." He may have wanted to abolish democracy and create a system in which America had a CEO rather than a president, but at least his literary references were superior to Thiel's.)

As he drove, Anthony talked of how he'd come to see Thiel as a representative figure of our time. He was feeling his way toward a kind of grand unified system he'd tentatively started referring to as "Thielism." This had arisen out of Silicon Valley libertarianism, he said, and encompassed a range of convictions about technology and the human future. A belief in monopoly capitalism. The mining and exploitation of personal data. Radical extension of life spans through technological means. Cryptocurrency as a method of evading both government regulation and the taxation on which nation-states depended. Above all, a belief in the emergence of a new "sovereign individual."

"It's about radical individualism," he explained. "It's survival-of-the-fittest, a belief in the rights of the wealthiest and most powerful among us to do whatever the fuck they want, including living forever. Thielism doesn't necessarily represent a human apocalypse, per se. Humanity could still continue living under its conditions. But it's an assault on the civilizational values I hold most dear, like creativity, empathy, love, freedom of expression, connection."

In a café in Queenstown, about an hour's drive from Thiel's estate, we met a man we'd been put in touch with by a wealthy and connected art world acquaintance of Anthony's. A well-known professional in Queenstown, he agreed to speak anonymously for fear of making himself unpopular among local business leaders and friends in the tourism trade. He had been concerned for a while now, he said, about the effects on the area of wealthy foreigners buying up huge tracts of land. ("Once you start pissing in the hand basin, where are you gonna wash your face?" as he put it, in what I assumed was a purely rhetorical formulation.) He told us of one wealthy American of his acquaintance, "pretty left-of-center," who had bought land down here to allay his apocalyptic fears in the immediate aftermath of Trump's election. Another couple he knew of, a pair of Bitcoin billionaires, had bought a large lake-side estate on which they were constructing a gigantic bunker.

This was the first I'd heard since coming here of an actual bunker being built. From the point of view of the modern apocalypticist, the whole appeal of the country—its remoteness and stability, its abundant clean water, its vast and lovely reaches of unpeopled land—seemed to be that it was itself a kind of reinforced geopolitical shelter, way down there at the bottom of the world. If wealthy foreigners were buying land here and building literal bunkers, fortifications beneath the ground of this country that had welcomed them in the first place, what did that say about their motivations, their view of life?

More than a year after my trip to New Zealand, the country was once again a focus of international attention, when an Australian white supremacist walked into a Christchurch mosque during Friday prayers and murdered more than fifty people with an assault rifle, streaming the killing live on Facebook. On several occasions that week, I found myself watching videos online of New Zealanders—both Māori and Pākehā—performing the haka for the Muslim victims of racist violence. The raw masculinity and aggression of the Māori war dance channeled into a gesture of inclusivity and love was something I found deeply moving. More than once, I was brought close to tears. I mentioned this to Anthony in a text exchange after the massacre, and he talked about how extraordinary the public response had been. The day after the attack, he and his family had gone to their local mosque, he said, to bring flowers and pay their respects, and the place had been filled with white families like his own, most of whom had never been in a mosque in their lives.

If civilization meant anything at all, I thought, it was this.

It was non-Muslim families crowding into a mosque the day after an act of fascist terrorism. It was a group of Māori men performing a war dance in the name of inclusion and solidarity and collective grieving: the precise symbolic opposite of fascism. It was not the building of bunkers beneath private land that would allow us to survive the catastrophes we faced, but the strengthening of communities that already existed.

In Queenstown, before we set out to find the former sheep station Thiel had bought, we went to look for the house he owned in the town itself. This place, we speculated, must have been purchased as a kind of apocalyptic pied-à-terre: somewhere he could base himself, maybe, while whatever construction he had planned for the sheep station was under way. We found it easily enough, not far from the center of town, and recognized it right away from one of the paintings in *The Founder's Paradox*. It was the sort of house a Bond villain might build if for some reason he'd been forced to move to the suburbs: ostentatious in a modest sort of way. The front of the building was one giant window, a glazed eye staring blankly at the town and the lake below, a home befitting a billionaire in the business of surveillance technology.

There was some construction going on in the place. I wandered up the drive and asked the builders if they knew who their client was. "No idea, mate," they said. They were just doing some renovation on contract. There'd been a fire in the place a while back, apparently. Nothing sinister, just a wiring issue.

The next day, we made our way to Lake Wanaka, where the larger rural property was located. We rented bikes in the town and followed the trail around the southern shore of the lake. It got rockier and more mountainous the farther we pursued it, and by the time we knew for certain we were on Thiel's property, I was so hot and exhausted that all I could think to do was plunge into the lake to cool off. I asked Anthony whether he thought the water was safe to drink, and he said he was sure of it, given that its purity and its plenty were a major reason a billionaire hedging against the collapse of civilization would want to buy land there in the first place. I swam out farther into what I had come to think of as Thiel's apocalypse lake, and, submerging my face, I drank so deeply that Anthony joked he could see the water level plunging downward by degrees. In truth, I drank well beyond the point of quenching any literal thirst. In a way that felt absurd and juvenile, and also weirdly and sincerely satisfying, I was drinking apocalypse water, symbolically reclaiming it for the 99 percent. If in that moment I could have drained Lake Wanaka just to fuck up Thiel's end-of-the-world contingency plan, I might well have done so.

I suggested I might take a rock, a piece of the place to bring home and keep on my desk, but Anthony warned me that to do so would be a transgression of the Māori understanding of the land's communal sacredness. We scrabbled up the stony flank of a hill and sat for a while looking out over the calm surface of the lake to the distant snowy peaks, and over the green and undulating fields unfurling into the western distance, all of it the legal possession of a man who had designs on owning a country, who believed that freedom was incompatible with democracy.

Later, we made our way to the far side of the property, bordering the road, where we saw the only actual structure on the entire property: a hay barn. It is the opinion of this observer that Thiel himself had no hand in its construction.

"There you have it," said Anthony. "Eyeball evidence that Thiel is stockpiling hay for the collapse of civilization." I wish to state categorically that we did not steal so much as a single straw from that barn.

We had made it to the center of the labyrinth, but it was elsewhere in the end that our monster materialized. In early December, a couple of weeks after I'd left the country, Max Harris, the young Kiwi author whose book Simon and Anthony had used as a counterpoint to Thiel's ideas, was home for Christmas and went along to the gallery to see the exhibition.

Down in the basement, in the central chamber—with its low ceilings, its iron vault door, its Führerbunkerishly oppressive vibe—Harris encountered, staring intently downward into the glass case containing the *Founders* game, a man in shorts and a blue polo shirt, surrounded by a group of younger men, likewise polo-shirted. The older man was doughier and less healthy-looking than he appeared in photographs, Harris told me, but he had little doubt as to his identity.

Harris, who was aware that Peter Thiel had not been seen in New Zealand since 2011, asked the man whether he was who he thought he was. The man smirked and, without raising his eyes from the board game toward Harris, replied that a lot of people had been asking him just that question. Harris asked the man what he thought of the exhibition, and the man paused a long time before saying that it was "actually a work of phenomenal detail." He asked Harris if he knew the artist, and Harris said that he did, that he himself was in fact a writer whose work had formed part of the conceptual framework for the show. Of the sheer improbability of these two men—one for whom New Zealand was a means of shoring up his wealth and power in a coming civilizational collapse, one for whom it was home, a source of hope for a more equal and democratic society—just happening to cross paths at an art exhibition loosely structured around the binary opposition of their political views no mention was made, and they went their separate ways.

Thiel left his contact details with the gallery, suggesting that Simon get in touch. He did, and Thiel responded quickly: he'd been intrigued by what he had seen but claimed to be a little disturbed by how dark his cyber-libertarianism appeared when refracted through the lens of *The Founder's Paradox.* In any case, the conversation continued, and they made arrangements to meet on Simon's next trip to the United States.

Simon was eager to keep talking, if only because he was determined to reach a deeper understanding of Thiel's vision of the future. Anthony, the more straightforwardly political in his antagonism toward Thiel and what he represented, was bewildered by this unexpected turn of events, though strangely thrilled by it, too. For my part, this came as a disorienting rug-pull ending—partly because the monster had materialized, and he was therefore no longer merely a human emblem of the moral vortex at the center of the market, but also an actual human, goofily got up in polo shirt and shorts, sweating in the heat, traipsing along to an art gallery to indulge his human curiosity about what the art world thought of his notoriously weird and extreme politics. A sovereign individual in the same physical environment as us ordinary subject citizens. But it also deepened the mystery of what Thiel had planned for New Zealand, for the future.

There was one mystery that did get solved, though not by me: the admittedly frivolous enigma of what sort of renovations those builders were working on at the apocalyptic pied-à-terre in Queenstown. Nippert, in a recent *New Zealand Herald* article, had published the architect's plans for the place. Thiel was making some alterations to the master bedroom. He was putting in a panic room.

## 5

# OFF-WORLD COLONY

Toward the end of the final episode of the National Geographic documentary series *Mars,* there is a scene where Elon Musk, the founder of the private space transportation company SpaceX, visits Cape Canaveral with his young son. Together they ascend the elevator up the launch tower to where the space shuttles once began their trajectory into space, and he explains to the boy that in years past this was exactly how the astronauts themselves would have ascended before launching. As Musk and his son look out over the Kennedy Space Center, the green swampland, the Atlantic Ocean beyond, the billionaire speaks in voice-over of his company's mission to colonize Mars, and of how it had always seemed to him that we should have gone there by now, that we somehow lost our way. "And now," he says, "we're going to get back there."

The series ends with a scene of the first successful launch of SpaceX's reusable rocket, a crucial aspect of the company's plan for establishing a colony on Mars. There is the sublime vision of the rocket, spreading its dorsal fins, positioning itself upright over the landing pad, coming to a miraculous rest amid a great torrent of flames. And we see in tight close-up the youthful faces of SpaceX's employees, refulgent in the hopeful glow of the rocket's fire, and an explosion of ecstasy and relief, as though, six years after NASA retired the space shuttle program and almost fifty after the last man walked on the moon, the world itself had been redeemed, the future restored to its rightful grandeur.

"It's kind of amazing," says Musk, "that this window of opportunity is open for life to go beyond Earth. And we just don't know how long the window is going to be open. The thing that gets me most fired up is that creating a self-sustaining civilization on Mars would be the greatest adventure ever in human history. It would be so exciting to wake up in the morning and think that that's what's happening."

I will acknowledge that I held Musk in more or less unwavering contempt, because as a union-busting billionaire who had hijacked the language of collective hope and aspiration to promote a private enterprise for sending wealthy people to Mars, he seemed to me to reflect what was most degraded and abject about our time. This was a man, after all, who had earlier that year attempted to win hearts and minds by launching one of his own Tesla cars into space. It was hard to imagine anything more pathetic, anything more stupid or crass, than launching a red sports car into perpetual solar orbit, sullying the vast inhuman emptiness of the cosmos with the shimmering trash of consumerism. (At the time, I wondered why no one at SpaceX had thought to point out to Musk that he could have ensured his red sports car would orbit the sun indefinitely by leaving it parked in front of his house.) I had this much in common with Musk's most fervent fans: I saw him as a mythical figure. But the myth I had in mind was one of my own imagining, in which a perfect simpleton had, out of some unknowable Olympian whim, been singled out by the gods and granted the threefold gifts of intelligence, ingenuity, and money, which gifts he employed in precisely the manner a simpleton would: to pursue a civilization on Mars, for instance, and to launch luxury consumer goods into outer space with a rocket.

And yet, as I watched these scenes on my laptop on a flight from Dublin to Los Angeles, I found myself strangely moved. Perhaps the low oxygen level in the cabin was affecting my cognitive functioning, but I felt that there was something poignant about this nostalgia for the future, this insistence that what seemed to have been irredeemably lost might in fact be regained. I could see why people were so deeply invested in Musk and his project, why they were so exercised and even moved by his presence. He seemed to me, in that moment, neither a divinely gifted moron nor a vicious robber baron, but rather a child, an innocent boy who wanted nothing but to go to Mars, and to inspire others to want to go with him.

The series ends with an inspirational speech from one Robert Zubrin, a former aerospace engineer at the defense contractor Lockheed Martin who in 1998 had founded an organization called the Mars Society, which advocated for the human settlement of Mars. Leaning toward the camera, eyes wide with a kind of anguished joy, Zubrin whispers urgently as though trying to awaken the viewer, the world, from a deep sleep: "Look up! Look up! There is everything out there!

There's trillions of other Earths! That's why we're going to do it. And the next time we go, we're going to go to stay." Watching this scene, I found myself overtaken by a kind of ecstatic melancholy. I watched it over and over, clicking rewind again and again, submitting to the strange feeling it provoked in me, a sadness that could not remotely have been intended, and which I myself barely understood. The phrase *trillions of other Earths* had an especially powerful effect on me, perhaps because it brought to mind St. John's words toward the end of the Book of Revelation: "And I saw a new heaven and a new earth: for the first heaven and the first earth were passed away; and there was no more sea."

It was Zubrin and the Mars Society that had drawn me to Los Angeles in that late summer of 2018, at a time when the worst wildfires in California history seemed at last to be submitting to a weeks-long containment effort. I was attending the Mars Society's twenty-first annual gathering at the Pasadena Convention Center. The organization had thousands of members, and chapters in twenty-eight countries, and it acted as both a public outreach organization and a political lobbying group for the cause of Mars settlement. It also maintained two research stations, one in the Utah desert and another in the Arctic, where teams of would-be pilgrims were dispatched for two weeks at a time "to simulate life on the Martian surface."

Entitled "Mars and the Space Revolution," the conference promised to explore how we might go about building a self-sustaining civilization on Mars. Around this central question, four full days of talks were scheduled across an array of topics. How quickly could a Mars colony become completely Earthindependent? What might a new Martian religion look like? How would a Martian colony be structured politically? How might the blockchain facilitate an interplanetary financial system? How could self-replicating robots be used to terraform a hostile alien environment? What were the logistics of drilling for water on Mars? There were numerous talks on the kinds of difficulties colonists might face on Mars, from natural disasters to teenage delinquency to the lack of a clearly defined legal regime for recognizing property rights in space under current US and international law. There was a talk by a Lutheran bishop entitled "Is Mars Exploration Virtuous?" (Given that the Lutheran bishop was also a founding member of the Mars Society, I felt confident in predicting that the answer would be yes.)

All these questions were in themselves interesting, but what I really wanted to know was where this fixation on colonizing Mars arose from, what it revealed about our relationship with the future of our own planet. I had long been of the opinion that there was no more lurid symptom of our current cultural malaise than the notion that we needed a "backup planet" for humanity. Although its advocates spoke of it as a manifestation of an indomitable spirit of exploration and adventure, it seemed to me to represent something like the opposite: an absolute surrender to an exhaustion in the bones of civilization. Since my trip to New Zealand—since my encounter with the site of Peter Thiel's planned apocalypse retreat, and with the logic of escape and conquest represented in Simon Denny's *Founders* board game piece—my fascination with the idea of Mars colonization had grown, and merged with my larger anxieties about an inhuman future.

Baldly stated, the idea was this: sooner or later, whether because of climate change or asteroid impact or some other unforeseen cosmic or terrestrial snarl-up, our planet would become utterly inhospitable to life. In order to avoid the complete annihilation of our species, therefore, we would by that point need to have established a human settlement elsewhere in the universe. Stephen Hawking, who in the final years of his life was one of the great secular prophets of apocalypse, put it as follows: "I am convinced that humans need to leave Earth and make a new home on another planet. To stay risks annihilation. It could be an asteroid hitting the earth. It could be a new virus, climate change, nuclear war, or artificial intelligence gone rogue. For humans to survive I believe we must have the preparations in place within one hundred years."

In his opening keynote on the first morning of the conference, Zubrin—mid-sixties, scholarly, yet with an air of squinting resilience—spoke of a space-flight revolution led by Musk, whom he portrayed as a redemptive figure in our darkened time. Even if he were to fail at this point, said Zubrin, he would still have succeeded, because he had proven beyond all reasonable doubt that it was possible for an entrepreneur, a private citizen, to do what only governments had previously been thought capable of. Creative forces had been unleashed, he said, and it was now clear how we could get to Mars.

It was Zubrin who had brought Musk into the Martian fold—before he started SpaceX, Musk had donated $100,000 to help fund the Utah station—and he was now a sort of John the Baptist with respect to the billionaire space entrepreneur. (Musk was one of a handful of tech billionaires, including Jeff Bezos and Richard Branson, who were investing large amounts of their fortunes in the prospect of privatized space travel, projects that were as often as not presented as a means of securing the future itself, as though the last hope for the species was the largesse of billionaires who possessed both the genius and heroism of spirit to save an imperiled humanity.)

Zubrin then proceeded directly to confrontation with the eschatological zeitgeist. He did not believe, he said, that we were living at the end of history, but rather at the beginning; and neither were we at the end of science, but at the beginning of that, too. We humans had certainly done okay so far, he said; we had gotten "the overture" done—getting out of the African savannah, peopling the farthest reaches of our home planet, building what he called a "Type One" civilization—but now it was time to begin the real work, the work of building a "Type Two" civilization. It was time that we became a space-faring species. And from there, he said, we would construct a much more potent humanity, a "Type Three" civilization that would spread into the outer realms of the galaxy, even the universe.

Just then my view of the podium was obscured by an elderly couple, latecomers, establishing themselves in the row ahead of me. The woman was especially ancient, proceeding with the aid of a walker, and wearing despite the heat of a Southern Californian morning a pair of powder blue woolen gloves, presumably as a measure against some or other painful skin condition. As her companion helped her into a seat and positioned her walker neatly in the aisle, I thought how strange it was for someone so old to be attending an event so resolutely oriented toward the future. Looking around the auditorium, I was struck by the general agedness of the assembled Mars enthusiasts. Perhaps it had to do with it being a Thursday morning, a time of the week more amenable to retirees (and writers of literary nonfiction) than to the typical gainfully employed Angeleno, but the average age seemed to me to hover somewhere around the mid-sixties. And it was impossible to overlook the overwhelming whiteness of the room: of the perhaps two hundred or so Earthlings who'd shown up for Zubrin's keynote, only one as far as I could see was black, and he happened to be stationed behind the video camera set up at the rear of the room—here, I surmised, for professional reasons, rather than any deep personal enthusiasm for colonizing distant worlds.

I kept hearing this word, *colonizing*, and it seemed to me a strange and revealing choice of terminology, given the significant weight of historical baggage attached to the whole project of colonialism (conquest, slavery, mass murder, subjugation, and so on). But the prospect of building a human civilization on Mars had a deeper motivation than that of ensuring the survival of the species: it was a fantasy of retrieving the idea of the future from the past, recuperating a twentieth-century optimism and excitement about technology and science, and rehabilitating it for the present. It was, in this sense, an exercise in future-nostalgia.

I had recently read Ashlee Vance's authorized biography of Musk, and this yearning for an age of colonial expansion ran through its pages like a hot shiver. Around the time of his first encounter with Zubrin and the Mars Society, Musk had logged onto NASA's website and had been appalled to find no detailed plan or timeline for the exploration of Mars. He was of the opinion, Vance writes, "that the very idea of America was intertwined with humanity's desire to explore. He found it sad that the American agency tasked with doing audacious things in space and exploring new frontiers as its mission seemed to have no serious interest in investigating Mars at all. The spirit of Manifest Destiny had been deflated or maybe even come to a depressing end, and hardly anyone seemed to care."

At the 2012 Mars Society Convention—in the very room, in fact, where I was now sitting—Musk had received a "Mars Pioneer Award" from Zubrin and had given a speech in which he explicitly linked a future of Mars exploration to an American history of colonial expansion. "The United States is a distillation of the human spirit of exploration," he said. "Almost everyone came here from somewhere else. You couldn't ask for a group of people that are more interested in exploring the frontier." (Musk did not allude in his speech to those who had been brought here against their will, or who had been here long before the frontier explorers he was invoking. What he meant by the "human spirit of exploration" was, in essence, the white European spirit of colonial conquest and exploitation.)

When Americans talked about settling Mars, it seemed to me that what they were really talking about was reinventing America itself, renovating the belief in their country's greatness not as mere reality, but as fable: as a morally instructive narrative of ingenuity and righteousness. Musk himself was not, technically speaking, an American—he was from South Africa, itself a kind of inverted United States, in which the minority project of colonial white supremacism had eventually been overturned—but I would argue that those who are most profoundly and indivisibly American are in fact those immigrants who are energized by a romantic understanding of the country and its foundational mythos of liberty and possibility. Americans are made, not born.

Zubrin was approaching the stirring final moments of his keynote. He was speaking of how, after a new civilization was established on Mars, we could then proceed to building new settlements on asteroids. There would be thousands of new worlds, he said, in which nonconforming people, people with different ideas of how human society should be organized, and who were therefore not popular back on Earth, would have a chance to build societies around their ideas. Many would fail, he said, but some would surely succeed, and they would show the way for the rest of us. So if we did what we could do in our time, he said—disperse the fog, spread the vision, establish humanity on Mars—then five hundred years from now there would be human civilizations on thousands of worlds in our solar system and others, and they would be as grand compared with what we are today as we are to our distant ancestors on the African savannah. Because we were not native to this Earth, he said. We were native to Kenya, which was why we had these thin arms with no fur. We could not have settled in North America or Asia if we had not developed technology. But we had, he said, because we were creative, and we were resilient, and that was why we were going to inherit the stars.

There was a long interlude of enthusiastic applause, during which I considered anew how intimately this rhetoric—Zubrin's talk of asteroid settlements, Musk's talk of the spirit of exploration—was entangled with the rhetoric on which America itself was founded: the apocalyptic invocation of the passing of an old world, the birth of a new. To speak of nonconforming people building new societies, of the entrepreneurial spirit of nation-building, was to explicitly appeal to an American mythology of pilgrims, founders, pioneers. Zubrin's Mars struck me as a futurist vision of the "city on a hill" invoked by the Puritan preacher John Winthrop in his famous speech to the passengers of the *Arabella* as they set out for the New World. Mars was America, I thought. The future was the past.

In his book *How We'll Live on Mars,* Stephen L. Petranek writes that "Mars will become the new frontier, the new hope, and the new destiny for millions of earthlings who will do almost anything to seize the opportunities waiting on the Red Planet." Like the first European colonists in America, the first humans on Mars, he says, will need to be extremely resilient and determined. This new New World, like its predecessor, will be profoundly hostile to settlement. They will need to find ways to make the air breathable, and to extract sufficient ice from the regolith, the Martian surface soil, to provide water. They will need to construct shelters, perhaps from regolith bricks, to protect them from the extreme cold and from the sun's radiation, which passes unfiltered through the planet's thin atmosphere. The example set by these pioneers, he writes, "will create a wave of fortune seekers to rival those of the California gold rush."

And just as the first European settlers in America saw themselves as ensuring the survival of Christendom, these first settlers on Mars will represent an insurance policy for civilization, for humanity itself. "There are real threats to the continuation of the human race on Earth," writes Petranek, "including our failure to save the home planet from ecological destruction and the possibility of nuclear war. Collision with a single asteroid could eliminate most life, and eventually our own sun will enlarge and destroy Earth. Long before that happens, we must become a spacefaring species, capable of living not only on another planet but ultimately in other solar systems. The first humans who emigrate to Mars are our best hope for the survival of our species."

Mars, as Musk once put it, is the "backup" planet for humanity, "just in case something goes wrong with Earth." But it represents something else, too, an idea much deeper and stranger and more difficult to sell. It is a means by which we—or certain of us, at any rate, with the will and financial means to do so—might leave behind our planet of origin, transcend the human world entirely. As with the imagined collapse scenarios of the doomsday preppers, Mars colonization is apocalyptic scenario as escapist fantasy. In her prologue to *The Human Condition,* Hannah Arendt writes about the response in the American press to the 1957 launch of the Soviet space satellite

*Sputnik,* the first human object ever to leave the planet and enter outer space. Notwithstanding the Cold War complexities of the launch, she observes, the immediate reaction was one of joy. But it was less a triumphant than a relieved joy—a "relief about the first 'step toward escape from men's imprisonment to the earth.' " (This statement, culled from a newspaper report on the event, was not merely the overenthusiastic framing of an American reporter, but in fact echoes the words etched on the tomb of Konstantin Tsiolkovsky, the Russian aeronautical engineer and space flight pioneer: "Mankind will not remain bound to the earth forever.")

The banality of the statement, Arendt insists,

> should not make us overlook how extraordinary in fact it was; for although Christians have spoken of the earth as a vale of tears and philosophers have looked upon their body as a prison of mind or soul, nobody in the history of mankind has ever conceived of the earth as a prison for men's bodies or shown such eagerness to go literally from here to the moon. Should the emancipation and secularization of the modern age, which began with a turning-away, not necessarily from God, but from a god who was the Father of men in heaven, end with an even more fateful repudiation of an Earth who was the Mother of all living creatures under the sky?

Reading Arendt's words, I hear in my mind the plaintive machine of Stephen Hawking's voice, narrating the BBC documentary *Expedition New Earth:* "We are the first species that has the potential to escape Earth." Like Musk and Zubrin, what Hawking is appealing to is a yearning for transcendence. There is, yes, an apocalypse that may happen—a man-made apocalypse like climate change; a cosmic apocalypse like the impact of an asteroid—but this is on some level a cover story for a deeper impulse, a desire to be done with the world itself.

And there is something fundamentally male about this narrative of exit, of escape as a means toward the nobility of self-determination. The cultural critic Sarah Sharma has argued for an understanding of exit as an exercise of patriarchal power, "a privilege that occurs at the expense of cultivating and sustaining conditions of collective autonomy." It's a force that she places in opposition to the more traditionally maternal value of "care." The politics of exit are pursued, she insists, at the expense of a politics of care. "Care," she writes, "is that which responds to the uncompromisingly tethered nature of human dependency and the contingency of life, the mutual precariousness of the human condition. Women's exit is hardly ever on the table, given that women have historically been unable to choose when to leave or enter inequitable power relations, let alone enter and exit in a carefree manner."

The world, after all, requires attention. The world requires care. To borrow Arendt's terms, to repudiate the Earth—which is to say, the Mother—is to reject the imperative of care. Mars represents a conquering of new territory, and a leaving behind of the old: a self-determination for the few at the cost of collective autonomy. The days of this world are numbered. For those who are willing to escape it, a new life awaits.

The frontier rhetoric around Mars colonization—the

invocation of pioneers, pilgrims, Manifest Destiny—brings to mind for me the advertising blimp that hovers over the filthy neon hellscape of downtown Los Angeles in one of *Blade Runner*'s early scenes. A gigantic screen displays the messages "Best Future" and "Breathe Easy," while a voice blares from its speakers, addressing the acid rain–sodden subjects below: "A new life awaits you in the off-world colonies, the chance to begin again in a golden land of opportunity and adventure." The voice is a male one. It is confident and cheerful, and richly reassuring. It is the very voice of American capitalism itself.

We return, for now, to Earth: specifically to a windowless ground-floor room in Old Pasadena, where a man named Art Harman—slacks, navy blazer, gold buttons—was standing behind a podium. Art Harman was the founder of an organization called the Coalition to Save Manned Space Exploration. He was a former adviser to the Trump presidential campaign; a conservative policy wonk specializing in both the expansion of American business interests into outer space and the protection of America's borders on the surface of Earth.

His talk was entitled "Liberty in Space."

Mars, Art Harman was saying, was the new land—the new planet—of opportunity. Mars colonization, as such, was all about free enterprise. A slide appeared behind him, a monochrome etching of ox-drawn covered wagons moving across the American Southwest, pioneers in broad-brimmed hats leaning back on their elbows against the desert scrub.

"It's about the spirit of entrepreneurship and everything

that goes with it," he said. "The unknown. Adventure. This is what has always made people give up safe and prosperous lives to seek out what's over the next mountain. Go west, young man!"

There was something about the way those first European settlers were valorized by the proponents of Mars colonization, at a time in which migrants from countries suffering the effects of political violence and climate change were relentlessly villainized, that felt to me like an intimation of a future in which a tiny minority of obscenely wealthy people were free to colonize other planets, mine asteroids, escape the smoldering wreckage of the Earth, while the poor and the desperate would be made to seem like invading armies, barbarian hordes. Was this the future, this hardening of hearts against spectacles of mass suffering, mass death? Was this the end of the world, or how it would continue?

Another slide: a jubilant crowd on the Berlin Wall, waving German flags, behind them the Brandenburg Gate lit by fireworks, the fall of an evil empire. Art Harman stepped out from behind his podium, squared his shoulders, adjusted his cuffs with emphatic precision.

"I was there," he said, "when that wall came down. The people in that photo, for the first time in their lives, there's not a guy with a gun to their heads when they try to express themselves. That's what it's going to be like for the first people on Mars. We're here. We're free. You won't have the government, the EPA, saying you can't damage this or that endangered species. Not on Mars."

On the screen, the Berlin Wall gave way to an image of the

much-fetishized preamble to the United States Constitution. This right here, he said, was the endangered species we should be trying to protect.

"Here in America we have a self-perpetuating free society, because we know our rights. We can defend these rights ourselves. It's not like that in most countries. Most countries, rights are offered like candies to a kid, and they can be taken back. In the Soviet Union, only the elite had rights, the guy with the gun. On Mars, we're going to want to avoid that."

His words echoed, quite precisely, the sentiment of the speech with which Vice President Mike Pence had outlined the vision behind the launch of the so-called Space Force, the new sixth branch of the US military, emphasizing the need for the increased militarization and privatization of space. "While other nations increasingly possess the capability to operate in space, not all of them share our commitment to freedom, private property, and the rule of law. As we continue to carry American leadership in space, so also will we carry America's commitment to freedom into this new frontier." It was all there: freedom, property, the rule of law. The sacral image of the frontier, a backdrop to it all.

I wondered how it was that so many Americans—educated, intelligent Americans—seemed to genuinely believe this stuff. Where did it come from, this conviction that their country was somehow uniquely possessed of a divine spark of freedom, a national genius for personal liberty? The only thing that seemed to me to explain the conviction also fatally undermined it: the fact that from cradle to grave every American was subject to a relentless barrage of propaganda about the special freedom guaranteed them by their citizenship. The answer, of course, was history, or rather the cultural products of its relentless mythologization. It was easier, in the way that all things were, to internalize this message if you were white—like Art Harman, like me, like everyone else in the room—and therefore less likely to be exposed to its negation.

Earlier that morning, on my way to get breakfast in Silver Lake, my eye had been drawn downward toward a small brass plaque embedded in the sidewalk on the corner of Sunset Boulevard and Hyperion Avenue. "PRIVATE PROPERTY," it read. "Permission to pass over revocable at any time." I had stood there a long while, reading it over and over, marveling at the strangeness of the message, its small and insistent authority in the surface of the street. Though it was presumably meant in earnest, it seemed to me a strangely subversive thing, a nearly subliminal assertion of capitalism's blank refusal of any boundaries to its territorial expansion. Someone else's right to own the ground beneath my feet, the right to remove it from under me at any time: What sort of freedom was this? I thought of all those covered wagons in Art Harman's slide, all those white men moving out here to the golden land, the off-world colonies, to stake their claims to property and liberty. Permission to pass over revocable at any time.

Moving back behind his podium now, Harman suggested that the first Mars settlements might best be viewed in the manner of an early American colony. People had become indentured servants in return for their passage to America, he said, and something similar might work for the Martian colonies. The National Geographic series *Mars,* he explained, depicted colonies that were jointly administered by national governments and corporations. And so perhaps, he said, you might work and live at an Amazon colony, a SpaceX colony, and have your rights and freedoms determined by the rules those corporations set. Art Harman seemed to think that this would be a good thing, that if such beginnings—whereby dispossessed peasants and impoverished workers spent a number of years working off their debts to businessmen who'd arranged their passage—were good enough for America, they were good enough for Mars.

In the front row, an English-accented voice was raised to point out that the United States Constitution, for all its insistence on equality and freedom, did nothing to prevent the institution of slavery persisting for almost a century after its framing. Art then made what seemed to me to be the most recklessly revisionist statement I'd heard all morning.

"Well, yes," he said, with patient courtesy. "But the founding fathers didn't like that, and the constitution was changed."

In the seat directly ahead of me, a heavyset man with close-cropped gray hair raised his hand. He began to speak, in a lilting and somewhat supercilious Scandinavian accent, of how in western Europe there were many different kinds of democracies, some of which—for instance, Norway, where he was from—were parliamentary systems, where it was not necessary to impeach a leader if they turned out to be corrupt or incompetent, where they could be simply deposed through parliamentary votes. Governments in these countries, he pointed out, tended to be seen as having many serious responsibilities. Where he was from, he said, health care was viewed as a basic right.

An unseen person shifted heavily in the seat directly behind mine, emitting a sigh that was in fact more enraged than weary. Art Harman asserted that what he was interested in here, in this talk, was in countering tyranny on Mars—and that, in any case, we were running low on time.

The exchange was palpably about something other than Mars. But then that was the thing about Mars: when people talked about it, they were always simultaneously talking about something else. As often as not, when people were talking about Mars, they were talking about America. But then nobody was ever just talking about America, either, because America, famously, was not so much a country as an idea: it was, specifically, the idea of itself as being an idea. The topic was, in this way, always threatening to spiral off into the low Earth orbit of abstraction. Something like this point had been made decades ago by Carl Sagan, the beloved astronomer and host of the enduringly popular 1980s PBS show *Cosmos*. (Sagan was always being invoked by Mars enthusiasts—often, it seemed to me, as a kind of emotional shorthand for a childlike cosmic optimism that had been lost since the twilight of the space age.) "Mars," he said, "has become a kind of mythic arena onto which we've projected our earthly hopes and fears."

Later, I attended a presentation by a couple of academic pediatricians about the effects a Martian environment might have on the children of colonizers. If we were going to have a self-sustaining civilization on Mars, they argued, that would mean children being born there. It was one thing to talk about settlers, they said, those people who would choose to make the long and perilous journey from Earth to Mars, and to establish a livable environment there. But in talking about a *colony,* they said, you were immediately talking about procreation, about future generations of children, who were after all the whole point of settling Mars in the first place. And these would be children who had never asked to be born at all, let alone born on a cold and inhospitable rock tens of millions of miles from the planet every previous generation of their species had been born on.

And this, these pediatricians reasonably pointed out, might present some significant problems both for the children themselves and for the adults who would have to care for them. The atmosphere on Mars, the lack of sunlight and the low-gravity environment, they said, would cause significant health issues for the first generations of Martian children. Children's bones developed, their cartilage formed, in direct response to the demands of gravity. Mars's low-gravity environment would present significant problems for physiological development. (The astronauts on the International Space Station do regular exercises specifically designed to lower the risk of osteoporosis and muscular degeneration, but these astronauts are, crucially, not babies and toddlers.) Intensive physical therapy, they said, would be needed if these first Martian children were to survive on their home planet. And that was before you even started talking about children having to live their lives underground, to protect them from the planet's much higher levels of radiation, or the problems of extreme boredom and depression that might well arise.

Neither the people advocating for human settlement of Mars nor the pediatric profession was giving much thought to the problem of children, they argued, and this needed to change.

I remembered something my son had said a few weeks previously, at a science museum we had taken him to. We were looking at an underwhelming audiovisual exhibit called "Colonizing the Cosmos," about how the human settlement of Mars might be the ultimate means of ensuring the survival of the species. "I don't want to go to Mars," he had said. "It doesn't look nice."

He was right, I thought. It looked like a total shithole. But then again, this planet was no paradise either.

Why would we want children to be born on a distant planet, a place with an atmosphere of mostly carbon dioxide, and precious little gravity: a place for which their bodies and minds had not evolved? I knew the answer, of course: because we needed to ensure the future existence of the species, and it was to this fundamental imperative that the first generations of Martian children would be sacrificed. This necessity was taken as self-evident, beyond the scope of refutation. But was it so unacceptable that humanity should eventually run its course? Why was it so unthinkable that we ourselves—not necessarily tomorrow or the next day, but eventually—follow the same well-beaten trail toward oblivion as the dodo, the black rhinoceros, the passenger pigeon, the Javan tiger, the sea mink, the great auk, the Yangtze river dolphin, the monkey-faced bat, the Aru flying fox, and all the countless other species whom we ourselves had driven from the face of the Earth?

I thought of the fires that had been burning here in California all summer. I thought of the drought back home in Ireland, the strange heat. I thought of how my son had been in the world for five years, four of which had been the hottest in recorded history.

In their lists of reasons for establishing a "backup planet" for humanity, advocates for Mars colonization invariably included the prospect of climate change making the Earth unlivable. And yet even in the most dire projections of Earth's future, there was no suggestion it might ever become as hostile to life as Mars, a planet with essentially no atmosphere, and on which the surface radiation levels were one hundred times that of Earth.

Earlier that summer, the US government had published a five-hundred-page draft environmental impact statement, intended to justify its freezing of fuel efficiency controls. Embedded in the statement was an acknowledgment that, based on current climate trends, average global temperatures were likely to increase by four degrees Celsius by the end of this century, an increase widely understood as disastrous by climate scientists. The statement's position, though, was not that something needed to be done about this, but rather that regulating emissions for new cars was an essentially frivolous exercise, given that a coming climate catastrophe was already a given. This was something far uglier than the denial of the reality of climate change. This was an acknowledgment of its likely catastrophic effects, and an insistence that there was now no point in trying to mitigate them through government intervention; an argument that it was better at this point to continue the destruction unimpeded. Because there were still things to be bought and sold, money to be made. Because there was still time.

As I was contemplating the prospect of human extinction, I realized that I was staring at the back of a man seated directly in front of me. He was wearing a black T-shirt, across the back of which was written the web address "marscoin.org." (The neck and shoulders of this T-shirt were stippled with dandruff in such a way that it seemed to depict the cosmos itself, a heavenly firmament of flaked skin.) As discreetly as I could, I slipped my phone from my trouser pocket and entered the address into its browser, and I found myself on the website of a cryptocurrency founded by Mars Society members who had intended it to act both as a source of funding for Mars colonization projects and as the eventual colony's de facto currency. "Marscoin," I read, "is dedicated to supporting the colonization of Mars and other space-related projects intended to get humans living and thriving off of planet earth. Simply by using and investing in Marscoin, you are contributing to a serious bootstrapping effort to further a colony on Mars." There seemed to be a general consensus in Mars colonization circles that the financial system of the colonies would inevitably be based in some or other cryptocurrency. (That there was a high degree of crossover between the enthusiasts of human settlement of Mars and blockchain fundamentalists was not especially surprising, given that both were of disproportionate interest to the libertarian wing of the geek community.)

It was right out of *The Founder's Paradox,* this whole idea. I thought again of the world-building strategy board game, down in that dungeon-like basement of the gallery in Auckland, depicting successive levels of escape from a dying planet, with its democratic nation-states, until the player finally reached the anarcho-capitalist utopia of Mars. In his catalog essay for the show, Anthony had quoted an article by Thiel in which he'd said that, when it came to the matter of escape, the important question was one "of means, of how to escape not via politics but beyond it. Because there are no truly free places left in our world, I suspect that the mode for escape must involve some sort of new and hitherto untried process that leads us to some undiscovered country." The undiscovered country was the Internet, yes, but it was also New Zealand, and it was also space itself.

I remembered Anthony, driving our rental car toward Thiel's estate on Lake Wanaka, talking about how he didn't want his son to grow up in the future people like Thiel and Musk were working to construct.

*No truly free places left in our world.* The kind of freedom that was being invoked here was the freedom from government, which meant freedom from taxation and regulation, which in turn meant the freedom to act purely in one's own interest, without having to consider the interests of others—which seemed to me the most bloodless and decrepit conception of freedom imaginable. (It was surely no coincidence, I thought, that little was ever said about building "communities" on Mars: the concept of community involved thinking of other people as more than burdens, or resources to be exploited, or rational actors with whom you could trade.) The notion of escaping "beyond politics" was, in other words, itself inescapably political. It was a dream of dissolving all entanglements with, and obligations toward, other people. This amounted to nothing less, in my view, than the dissolving of life itself.

At the time of the convention, notwithstanding the occasional plane coming in low toward LAX, Mars was the brightest thing in the night sky above the city. Mars, the planet closest to our own, is no particular distance away. Because the two planets are elliptically orbiting the sun at different rates, the distance varies from 33.9 million miles at its shortest to 250 million miles at its longest. In the late summer of 2018, toward the end of that long and devastating fire season, Mars was closer to Earth—or, and this was somehow more unsettling to consider, Earth closer to it—than at any point in the previous fifteen years. If you were going to set out for Mars, or return from there to Earth, now would be the time.

"To be living here, and not in what we mostly believe is the insupportable there, elsewhere, is to be assimilated into a powerful abstraction, the abstraction of never-ending possibility," wrote Elizabeth Hardwick. "The American situation is not so much to overthrow the past as to overthrow the future before it arrives as a stasis."

The chaos and upheaval and entropy of our time, its roiling surface of radical change: Are these not in fact hysterical symptoms of a deep and lethal stasis? Everything is falling apart, coming to an end, precisely because we are unable to believe in the possibility of change. And what is true of the West in general is, as always, spectacularly, gruesomely true of America in particular. At the risk of stating the obvious: nobody is going to make America great again. Nobody even seriously imagines it to be a possibility. America might, it is true, eventually stop outsourcing its manufacturing to China, but if those jobs are ever brought back home, they will return in the form of automated labor. Robots and algorithms will not make America great again—unless by "America" you mean billionaires, and by "great" you mean even richer. Its middle class has been gutted, sold off for scrap. Trump is only the most visible symptom of a disease that has long been sickening the country's blood—a rapidly metastasizing tumor of inequality, hyper-militarism, racism, surveillance, and fear that we might as well go ahead and diagnose as terminal-stage capitalism.

What Hardwick calls the abstraction of never-ending possibility has its historical precedent in the frontier. Among the many other things it is animated by, abstract and concrete, America is animated by a foundational imperative of expansion. And this much it has in common with another of its great animating forces, capitalism, which exists and thrives through expansion of its own frontiers, through a relentless force of deterritorialization. And it is running out of frontiers; running out of boundaries to obliterate, nature to exploit. The legacy of its monomaniacal pursuit of cheap resources is a devastated planet that may soon be unlivable for vast numbers of its inhabitants.

"Human beings can't go west anymore," write Charles Wohlforth and Amanda R. Hendrix in their book *Beyond Earth: Our Path to a New Home in the Planets.* "Our planet is full. Our personality as a species suggests some of us won't put up with that situation indefinitely."

The fantasy of colonizing Mars is, to employ Hardwick's terms, a means of overthrowing the future before it arrives as stasis. In calling it a fantasy, my intention is not to dismiss the idea as a mere delusion. The world, after all, is built on mythology. The foundations of our flimsy reality rest in the bedrock of fiction.

Once you start using the apocalypse as a way of encountering the present, an anxious response to uncertainty and change, it presents itself everywhere in the form of cryptic signals, deep emanations. On the final afternoon of the convention, I took an Uber back to the city from Pasadena. I got talking to the driver, a man of about fifty whose name was Alexander, and he told me about his childhood in the Philippines. His parents had both died when he was very young, he said, and he'd been taken in by an older brother, who himself had many children and little time or inclination to look after him. And so he'd grown up on the streets of Manila. Somewhere along the way, he'd picked up a pretty serious gambling addiction, which had only worsened when he moved to California in his twenties.

But all that had changed now, he said. He hadn't held a deck of cards in years, felt the rattle of dice in his closed fist. Something in his manner, a kind of entrepreneurial approach to the retailing of his own story, told me I was talking to a born-again Christian, and indeed we had only just hit the freeway by the time he turned his attention toward his personal relationship with the Redeemer.

Between one thing and another, he got to talking about the end of the world, a subject that was drawn to me as much as I was drawn to it. The catastrophes that were happening now, he said, were so much worse than they ever had been before, and were happening so much faster. The world had gotten so bad. God had told us to love one another, and we were doing so much greed. God had said I gave you all enough to eat, enough to live good lives, but still there was so much suffering in the world. So many people doing greed, he said: everywhere greed. He was in no doubt, he told me, that he and I both would live to see the end. Noah's flood would come again in our time. Though not myself a Christian, or really much of anything at all, I was helpless to resist the lure of this talk, this language of floods and redemption, a personal God driven by love and vengeance. I didn't get the feeling he was trying to convert me, for what it was worth. My sense of it was we were just two guys talking.

Somewhere around Glendale, we stopped at some lights. I looked out the window and saw in front of a gas station an old man with long hair and a beard, yellowed and fulsome. He had headphones on, and he was jogging rhythmically on the spot. His eyes were closed and he was grinning with startling abandon, and in his hands he held a white wooden cross, not quite to scale, which he jounced up and down as he jogged. He was a sight to see, though nobody was paying him the least attention.

I remembered the kid I had seen by the river in South Dakota, wilding out with his headphones on. He, too, was in some trance of rapturous communion, impenetrable to the casual witness, comprehensible only as madness. I imagined that these two men were somehow connected, that they were members of a secret shamanic order, some vast collective of lone ecstatic dancers, channeling the death-seeking energies of the culture, the historical poison in the soil.

I remembered, then, something I had read not long before about a cultic ritual known as the Ghost Dance that had become widespread among certain Native American tribes toward the end of the nineteenth century. This was a circular dance, undertaken for many hours at a stretch, whose practitioners believed that in so doing they were hastening the end of the world, that the dance itself would cause the ground to open up and consume the colonizers and all their works, and that during this time the Native people would be ascended to heaven, to be replaced safely on their land at the end of the time of destruction.

The dance, needless to say, did not bring about the desired result. From a Native point of view, the apocalypse was in any case a matter of historical record.

# 6

# UNDER THE HIDE

When he was four, my son got very heavily into Dr. Seuss. By this I don't just mean that he enjoyed his mother and me reading the books to him—although there was certainly that: a great many bedtime giggling fits were sustained by way of exposure to the morally questionable high jinks of *The Cat in the Hat,* the alliterative absurdities of *Fox in Socks*—but that he was so affected by the books that he began to think about their author in a way that seemed to me to mark a deepening of his engagement with the world. What I'm saying, I suppose, is that his discovery of Dr. Seuss was also his introduction to the concept of the artist: a recognition of the fact that a book or a song comes to exist in a different sort of way to a Lego Minifigure or a chocolate bar; that it is the creation of a particular person with an irreducible experience and manner of expressing it.

It was agreed that Dr. Seuss was a genius. It was further agreed that Seuss's magnum opus was *The Lorax.* Certainly, it was the book that provoked the most questions and conversations, whose language and ideas were most thoroughly absorbed into our everyday exchanges. If it's not an outright work of postapocalyptic nonfiction, then *The Lorax* is about as close to being one as an illustrated book for preschoolers has any business being. Its setting is a bare and blasted landscape, where nothing grows save for a spindly black weed called "Grickle-grass," and it's in this dead place—where all birdsong has fallen silent save for the occasional croaking crow, and where the "wind smells slow-and-sour when it blows"—that we encounter a mysterious and sinister character named the Once-ler, who dwells in an impossibly rickety tower with boarded-up windows. The Once-ler's face is never revealed to us in the illustrations. Seuss represents his villain-protagonist as a pair of long green arms and, now and then, two yellow eyes staring eerily from the crepuscular recesses of his tower. We meet this uncannily partial personage via a character clearly intended as a stand-in for the child to whom the story is being told: a little boy who is in fact explicitly referred to as "You" and who, as the book begins, has tracked down the Once-ler to his hideout "at the far edge of town" in order to find out what happened to an even more mysterious character known as the Lorax.

Once the Once-ler has been remunerated for his narrative services (fifteen cents, a nail, a snail shell: because even in the wake of complete economic and ecological collapse, there is still a buck to be made), he begins to tell his story. He takes us back to an Edenic world of natural abundance and beauty, a lush landscape of soft and tufty Truffula trees, green rolling plains, and beatifically smiling animals of various Seussian species. Into this prelapsarian realm the Once-ler arrives—

rendered, again, only as a pair of green arms. Upon arrival in this place, the Once-ler informs us, he is taken above all by the beauty of the Truffula trees, in which he immediately sees an opportunity for wealth creation. He builds a little shop, and then chops down his first Truffula tree, using its tuft as a textile from which he knits a garment called a Thneed, which is ugly and ungainly to the point of outright absurdity.

It is then that we meet the Lorax, a stout and extravagantly mustachioed creature who looks somewhat like the actor Wilford Brimley rendered as a soft toy. He emerges from the stump of the first felled tree, to take vocal issue with the Once-ler's mistreatment of the land. ("I am the Lorax," he says. "I speak for the trees.") It's the Thneed itself, though, that the Lorax is particularly bent out of shape about. What even is it, he demands to know? What conceivable purpose could such a haphazard accoutrement serve, and why on earth was it deemed worthy of chopping down a beautiful tree?

The Once-ler then patiently explains that this Thneed is "a Fine-Something-That-All-People-Need"—by which he means that it's a shirt or a sock or a glove or a hat, or a carpet, or a pillow, or a sheet, or a curtain, or indeed a cover for a bicycle seat. The Lorax, speaking now for the reader as well as the trees, accuses the Once-ler of having lost his mind with greed, insisting that no one is ever going to buy such a pointless piece of merchandise. But he's wrong: the demand for Thneeds quickly becomes so great that the trees can't be felled fast enough, and a new Super-Axe-Hacker is invented in order to cut down entire rows at a single blow, while belching great plumes of smoke into the air and destroying animal habitats.

The Once-ler allows that the situation is regrettable, but insists that it can't be avoided, arguing along familiarly circular lines: "Business," he says, "is business." And people, after all, want Thneeds.

This particular moment in the book has given rise to the occasional bedtime discussion on the nature of consumer desire, on where needs end and Thneeds begin. My son will point out that Thneeds are silly, and that the Once-ler's customers are themselves "total stupidheads" for buying them. I will agree but judiciously point out that we are all of us on occasion prone to lavishing money on the odd Thneed, that it's therefore worth bearing in mind the extent to which we are all of us, in our own right, total stupidheads, and that anyway it probably isn't nice to call people total stupidheads, though I take his point.

"We don't have any Thneeds," my son protests.

"That's technically true," I say. "Because Thneeds aren't a real thing, and you couldn't buy one even if you wanted to. But maybe what Dr. Seuss is getting at is the way we all tend to buy things we don't need. I think it's a metaphor."

At this point, I can hear the clanking and wheezing of the machinery of my PhD in English literature as it is roused into a state of sluggish animation.

"Do you know what a metaphor is?" I ask.

My son turns his face slightly toward the wall and literally tightens his lips, as he tends to do rather than admit to not knowing something. I have lately noticed in him the presence of intellectual vanity, and although I know it's not a particularly likable characteristic in people generally, I can't help finding it adorable in him, if occasionally frustrating. ("I know that!" he'll say crossly when told a thing he already knew, and occasionally a thing he didn't.)

To the best of my ability, I explain to him what a metaphor is, though I'm not sure he's really grasped it.

"So maybe a Thneed is basically anything we don't actually need," I say, "but really want anyway."

"Like what?" he says.

"Like maybe Lego Minifigures?" I suggest.

"No," he says. "Minifigures aren't Thneeds!"

He's riled now, leveraging against my shoulder to hoist himself upright in bed, and I'm wondering whether this discussion might ultimately be damaging to the larger objective of the bedtime story, which is, in theory at least, to wind him down for sleep.

"Aren't they?" I counter suavely, and more or less against my better judgment. For the last few months, my son had been very into Lego Minifigures. These were packaged in such a way that you never knew which figure you were going to get until you had bought and opened it, and so the more of them you bought, the more likely it became that the packet you'd just purchased would contain a Minifigure you already owned—a guy in a banana suit, a Lego Batman in a pink tutu, a zombie in business attire—thereby making the acquisition completely superfluous. Paradoxically, though, because the Minifigures came in limited edition series, an element of completism came into play, and the closer he got to having the full complement of a particular series, the more determined he became to get there, even if it meant the acquisition of duplicates or triplicates of particular figures. ("Ugh, the hotdog suit guy again!") On some level, he understood—because his mother and I had explained it to him—that this was a cynical marketing ploy on the part of the Lego corporation, but this understanding did nothing to dampen his appetite for further acquisitions. There was, at one point, a guy on staff in the toy shop we frequented who had an uncanny ability to divine which figure a package contained by means of lengthy and detailed manual palpation, but he eventually moved on—hopefully not because he was fired for using his God-given gift for divination to help customers like ourselves hack the Lego Minifigure system.

Offended by the implication that Lego Minifigures might be categorized as Thneeds, my son then delivered a shamelessly low blow by suggesting that coffee was a Thneed.

"I don't think coffee counts as a Thneed," I said.

"But you don't need coffee," he said.

"Well, yeah," I said, "nobody *needs* coffee, but it actually has a lot of health benefits."

"What's benefits?"

"A benefit is a thing that's helpful. A good thing."

"But you have too much of a good thing. You're only allowed"—and here he paused to calculate, more or less arbitrarily, how many cups of coffee I was to be permitted—"two cups of coffee a week."

"A *week*?"

"Four cups," he conceded.

"I thought I was allowed two cups per day."

"Four a week."

This was an ongoing argument, one that in truth I greatly enjoyed. Somehow he had sensed, in my admittedly overeager coffee consumption, a fast track to the moral high ground, a means by which he could legitimately critique me for the kind of overindulgence (in his case, chocolate and cakes and so on) he felt I unfairly chided him for.

It was agreed, eventually, that anything beyond two cups of coffee per day was to be considered a Thneed. He then suggested that books might also be considered Thneeds. For all that I received this as an outrageous claim, and an even lower blow than the coffee suggestion, it wasn't entirely out of left field. Two or three mornings most weeks, a package would arrive at the door of our house, and my son would make an elaborate show of exasperation at my receipt of yet another book. "More books? Too much of a good thing, Dada!"

"I really don't think we can say books are Thneeds," I said. "In that case am I a Thneed-maker? Is Dr. Seuss himself in the business of making Thneeds?"

He responded how he always responded when he knew I had him cornered in an argument, which was the same way he tended to respond when I'd made a joke that he knew to be a good one but was unwilling to grant me the satisfaction of laughing at: he groaned and shook his head. "Ach!" he said, and pursed his lips in an effort not to smile, and he gave me one of his "hard stares," a technique he'd picked up from the Paddington books. Returning a hard stare of my own, I thereby wordlessly initiated, via agreed protocol, a staring match. This was among my favorite games to play with my son, because it provided a useful pretext for gazing at leisure into his eyes, for a kind of intense perusal—leading to an almost unbearable refulgence of tenderness—for which in the ordinary run of things I pretty much never had occasion.

As fun as it is to read *The Lorax* to a child, it is also a sorrowful ritual. Because even though the child may laugh at the Lorax as he glances sadly backward at the Once-ler and lifts himself into the air by the seat of his pants—and even though you may laugh with him—you can't ignore the story you are ultimately telling him. The last Truffula tree is gone, and the Thneed factory has been shuttered, and nature itself is in its death throes. You are telling the story of the world he has been born into, and his likely future in it.

And then there is the ending, with its terrible gesture of hope. The Once-ler, in the book's final lines, returns to addressing his interlocutor, the little boy who stands for the child being read to, who stands for my son. He speaks of a small pile of rocks near the abandoned factory, whose large frontal stone is mysteriously engraved with the word "UNLESS." The Once-ler has long puzzled over the meaning of that word, but now that his interlocutor—now that my son—is here, he says, "the word of the Lorax" is no longer so mysterious. "UNLESS someone like you / cares a whole awful lot, / nothing is going to get better. / It's not."

The Once-ler throws something from the high window of his tower, instructing the child to catch it as it falls. It's a Truffula seed, he says, the very last one in existence. He instructs the child to plant it, and water it, and make sure that it gets clean air, and to grow from it an entire forest of Truffula trees. And perhaps then, he says, the Lorax and all of his friends may come back.

Strange as it may be to say this of a children's book, these closing lines are among the most heartbreaking I've ever read. The context has some bearing on this, obviously: they're heartbreaking because I'm reading them to my son, and because the message they convey carries what feels like a generational dead weight. The entrusting of the last seed, and the set of instructions for the revival of nature, must have been onerous enough when *The Lorax* was first published in 1972, in the very earliest days of the environmentalist movement—before we knew what we now know about climate change, before we understood that the world's sixth mass extinction event was under way in our time, and that we were the cause of it. Even in 1972, the imperative of those closing lines must have felt like an unbearable burden to place in the hands of a child. ("It's up to you! We, the previous generations, have fucked up everything, and only you, our innocent redeemer, can do what is needed to reverse the devastation of nature!") But now, it can feel downright cruel. Because what if it's too late now, and has been too late for a very long time? What if the Lorax and his friends are never coming back? What if the world my son has been born to, and will have to somehow make a life in, is one where nothing but Grickle-grass grows, and the wind smells slow-and-sour when it blows?

—

I thought about *The Lorax* a lot during the week I spent in Alladale. Not just because the questions I thought about when I was there were the questions raised by the book, but because the story of the place is also in some sense the story told by the book. It's a place that was once entirely covered by trees—oak, pine, ash, birch—that once forested the islands of Britain and Ireland. And there were wolves there once, and bears, too, but no longer: the animals left, died out, because the trees were all cut down. The Once-ler, in this particular case, was colonial expansion, the Industrial Revolution, the birth of capitalism.

To spend time in Alladale Wilderness Reserve was to calculate the intimate equation between civilization and environmental devastation. Though it was one of the last wild regions of the British Isles, this vast estate, twenty-three thousand acres of glacial valley and bare mountains in the Scottish Highlands, was in fact the private property of an Englishman named Paul Lister, the heir to a retail discount furniture fortune who purchased it in 2003 with the intention of reforesting the land and reintroducing several animal species—wolves and bears, most controversially, but also lynx and elk—that had become extinct in these islands as a consequence of the deforestation that fueled a history of colonial shipbuilding, industrialization, and retail discount furniture chains. And so while Alladale, in its remoteness and pristine stillness, gave the impression of being entirely innocent of human corruption, it was in fact a landscape stripped bare by the machinery of profit, a place of emptiness and silence: a place where the Grickle-grass grows.

It was out of sad curiosity about the dimensions of the loss we were undergoing as a civilization, as a species, that I decided to spend a week there in the spring of 2017. And I was not alone in this endeavor. There were sixteen of us on this retreat into the wilderness, total strangers who had come together with the intention of discussing the most pressing questions of our time, and of forming in the process a kind of tribe. This was not the sort of explicitly romantic endeavor I would ordinarily involve myself in, what with the unwieldy carapace of cynicism I had allowed to grow around me over the course of my adult life. There was some part of me that considered such pursuits fundamentally self-indulgent and even frivolous, and yet this was in conflict with a deeper sense that nothing could be more important—in, as it were, the end—than the unflinching engagement with the reality that we as a species might be finally and irrevocably fucked.

I had learned about the retreat on the website of an organization called the Dark Mountain Project, who were arranging it in collaboration with another group called Way of Nature, an English group that ran wilderness expeditions in locations around Europe. The Dark Mountain Project was a movement of artists and writers and activists united by the conviction that climate catastrophe was not just real and imminent but also in fact inevitable, a done deal, and that, as such, the entire project of environmentalism was effectively doomed. The group had been founded in 2009 by Paul Kingsnorth and Dougald Hine, two English former journalists and environmental activists who had begun to see the logic of environmentalism—its insistence that if we change our ways and embrace sustainable development and other eco-friendly practices, we might still prevent or mitigate the worst effects of climate change—as based on a delusion, as willful in its way as the delusion of those who deny the very existence of climate change.

The group's founders, along with its loose cohort of contributors and associates, were of the opinion that Western civilization was bound for collapse, that the whole vast edifice of interconnecting supply chains and technological infrastructures and political systems was more fragile than we permitted ourselves to recognize. "The pattern of ordinary life, in which so much stays the same from one day to the next, disguises the fragility of its fabric," as they put it in the manifesto with which they'd launched the group in 2009. The manifesto was written as the global financial crisis unfurled its devastating effects, and to read it is to sense its authors' dark and palpable exhilaration at the vulnerability of the systems undergirding our civilization. "There is a fall coming," they wrote. "We live in an age in which familiar restraints are being kicked away, and foundations snatched from under us. After a quarter century of complacency, in which we were invited to believe in bubbles that would never burst, prices that would never fall, the end of history, the crude repackaging of the triumphalism of Conrad's Victorian twilight, hubris has been reintroduced to nemesis."

The manifesto's central contention was that the foundational myth of our civilization—the myth of progress, the understanding of the future as a line on a graph that will soar ever upward and to the right—had been fatally undermined in our time. And this myth, it argued, was built on the foundations of a deeper myth: the myth of nature, the ancient idea that we, as a species, were fundamentally distinct from the world out of which we'd emerged. Climate change was both the most disastrous consequence of the belief in that myth, and the means of its destruction. It was climate change that brought us face-to-face with the instability of the civilization we had built, with how "the machine's need for permanent growth will require us to destroy ourselves in its name."

The manifesto was essentially an apocalyptic text, a revelation of what is undoubtedly to come, and in some sense a welcoming of its arrival. It rejected all suggestions that we might negotiate a means of evading this collapse—through political consensus, through technological ingenuity, through pursuing a more sustainable form of consumerism—and that everything might somehow turn out fine. As a species, it argued, we are in collective denial about our way of life and its long-term prospects for survival; even when we face up to the magnitude of the crisis we face, we tell ourselves it is merely a crisis—a difficult but soluble situation, rather than an approaching cataclysm.

Kingsnorth and Hine had drawn a lot of negative attention for the unremitting pessimism of their vision of the future, but what struck me most on reading it was its undercurrent of stern utopianism. As is the way of much apocalyptic writing, from John of Patmos to Karl Marx, the manifesto was animated by the desire for the immolation of a corrupted world, and the hope of witnessing a new dawn rising above its ashes. Beyond the uncompromising insistence that the rising tide of climate change would wash the Earth clean of our civilization and all its works, the Dark Mountain Project argued for a displacement of humanity from its seat as the center and source of all meaning in the world, and for an enactment of this displacement in new forms of "uncivilized" art and literature and storytelling. And what it ultimately gestured toward was something strangely hopeful, at least on its own terms: a world beyond the collapse of technological civilization in which humans—those humans who survived such a cataclysm—would find themselves no longer above or beyond nature, but within it, in a place where such categories as "human" and "nature" were no longer useful distinctions. "The end of the world as we know it," as they put it in the manifesto's closing lines, "is not the end of the world full stop. Together we will find the hope beyond hope, the paths which lead to the unknown world ahead of us."

On the evening of my arrival in Alladale, I had very real trouble getting my tent into any kind of tent-like configuration. The opaque disk of the sun was disappearing behind the great shadowing slopes of the western highlands, and the remaining light was dwindling inexorably, and I was, I realized, in serious danger of having to get the thing up in the dark, with only a headlamp to light my way, when a young woman who'd had her own nearby tent up for a while now poked her head through its entrance flap and asked whether I needed a hand. I will not pretend that accepting her offer did not involve a certain measure of masculinity-related discomfort, but it seemed to me that the embarrassment of politely declining, only to then continue foundering in tent-purgatory, would be exponentially more severe than the comparatively benign embarrassment of accepting.

"I used to work in a camping store," she said, "so I'm pretty used to putting tents up. They're tricky bastards, some of them."

I was immediately struck by the empathic skillfulness of this revelation, by the deft footing with which the woman had positioned herself not as a human being of normal competence, but as someone whose time as a camping store employee had endowed her with a special facility in tent construction—thereby skillfully maneuvering me out of my own position as an idiot who could not put up his own tent, and into the much less mortifying position of a person who, never having worked in a camping store, could be forgiven for not having been initiated into these esoteric practices. The profound emotional intelligence of this tactic was, in truth, even more impressive to me than the efficiency and speed with which she was putting up my tent in the dwindling light, which I knew was something any fool could do, myself excepted.

This woman's name was Amelia Featherstone. (It is only now that I'm writing about her, incidentally, that her name's amalgamation of emblematically opposed images—feather, stone—strikes me as somewhat heavy-handed in its paradoxical poeticism. This is hardly something Amelia herself could be blamed for, but neither is it something for which I can be held accountable.) As she put up my tent, and as I pretended to help her, she told me she was from Melbourne and that she worked for the government, in ecological conservation. This revelation prompted me to talk about how, as much as I wanted to see Australia, it was precisely its ecological diversity that gave me pause. Australia's sheer abundance of flying, skittering, crawling horrors constituted a deal-breaker, I said, in terms of my prospects for ever visiting the country.

I asked Amelia whether snakes were a particular concern in Melbourne, and she said that she did from time to time come across them in her work as a fire department volunteer.

"Now and then," she said, "you'll get a situation where you're dealing with a bush fire, and there are snakes in the bush, and the snakes are leaping out of the bush towards you."

"You mean, like, right at you?" I said.

"Yeah," she replied, in a tone that sounded to me almost apologetic.

"At your face?"

"More or less, yeah. Not deliberately, but you're in their path. And they're on fire, of course, when they're leaping at you, which is not great."

"So this is something that has happened to you, in your own life, as a person? Actual snakes that are on fire have leapt towards you from vegetation that is also on fire."

"Yeah," she said, and chuckled happily.

"I would not enjoy that at all," I said.

"Yeah, it's not great, as I said."

I was out of my comfort zone. It was a narrow zone, but deceptively spacious, and I did not like to be out of it. My comfort zone had good Wi-Fi and 3G coverage, and you could get

Japanese food delivered to it, and there was craft beer within walking distance, and bookshops, and it was clean and it was at all times more or less room temperature. It was a good place to be, my comfort zone. There were rarely spiders in it, and never any spiders that were on fire. There was not much nature in it at all, in fact, unless you counted potted plants, which were very much optional. My comfort zone was, strictly speaking, inside.

I had been thinking about this quite a bit before I came to Alladale, about my somewhat arm's-length relationship with nature. I was all for nature in theory, but in practice I had no feel for it, no sense of any relationship with it at all.

Actually that is not entirely true, because I was afraid of it—or certain aspects of it, at any rate, certain parts of it—and to be afraid of a thing is to have a relationship with it, however dysfunctional. I had certain quite intense nature-based phobias. I was, most pressingly, profoundly terrified of moths. It's a phobia I'd had for as long as I could remember, and was so mysterious to me in its weird urgency and intensity that I could only conceive of it as psychologically fundamental. It seemed to me that to disclose its origins would be in some sense to uncover the truth about myself.

When a moth enters a room I am in, or when I enter a room in which a moth is already established, it has long been my custom to swiftly withdraw. I cede the territory, no questions asked.

What is it about these small, defenseless creatures that so overwhelms me with elemental fear and disgust?

I find their blunt, furred bodies and twitching wings unpleasant to look at, certainly, but it is the manner of their movement that I find especially horrific: the total randomness of it, the indiscriminate courses of their flight. A moth will dart in one direction and then, for no good reason, just switch trajectories and double back the way it came. If your face happens to be positioned at any point along that trajectory, chances are the moth will blunder into it. And to be touched by such a thing, to have its body in contact with one's skin, seems to me a prospect beyond the realm of the thinkable.

I'd been seeing my therapist for more than a year by the time I brought this phobia up, and I could detect in her reaction some surprise that it had taken me so long to turn my attention to something so obviously ripe for the picking, so dense with analytic possibility. I had long thought of my moth phobia as an essentially comic neurosis, as a strange but basically minor personality quirk whose mysteriousness constituted a kind of psychological parlor game. (What could it mean? Where did it come from? Why *moths*?) But what I found as soon as I began to speak about it, there in her office, was that I wasn't finding it very amusing at all. I was finding it strangely difficult to talk about, as though I were suddenly on the precipice of some vertiginous chasm of significance. I was aware of a tension in my stomach, a tightness around something mutable and volatile.

We returned to the subject of moths repeatedly over a period of a few weeks, and it gradually emerged that this creature, this malevolently charged *object,* was linked via certain key associations to a general nexus of anxiety. I'd been talking about how I hadn't been sleeping all that well, about how I felt my sleep was more fragile than I'd like it to be.

*"Fragile,"* she said. "I've noticed that this is a word you use quite a lot. What comes to mind for you when you say this word?"

I said nothing for what seemed a very long time, attended to a couple of seagulls screeching at each other, the desolate staccato of a can being punted along the street by a gust of wind.

"Do you really want to know?" I said, straightening myself up on the couch.

"Of course I really want to know," she said, smiling.

"Death," I said. I was smiling back, but I was surprised to hear myself say it, and strangely chastened. I was feeling the same unease, the same tightness, that I had felt when I spoke about my moth phobia. "What I think about when I say the word *fragile* is death. The fragility of life. And the unpredictability of the future. It's the same thing. And I think this terror that I have of moths is also a terror of that fragility and unpredictability. Because they are so unpredictable. They are completely chaotic in their movements, and chaotic in their effects. I feel like my fear of moths must be some version of my fear of the future."

She took some slow and deep breaths, which I knew was her manner of encouraging me to do the same. We said nothing for a long time. Seagulls. Scraping can. Passing tram.

"Okay," she said. "Good."

We were out of time.

—

We were camping at Alladale, yes, but camping was by no means the sum total of what we were up to. What we were up to, as my wife had semi-ironically put it to me before I left for Scotland, was camping about the apocalypse. We were sitting around, cross-legged on riverbanks and reclined in clusters against grassy slopes, talking about the bad times we were in, the troubled days. What did we talk about? And who, in fact, were we, who were doing the talking?

We were a heterogeneous group: a handful of writers and artists; a recently retired business improvement district manager for a small town in Cornwall with an abiding interest in shamanism and other esoteric practices; a solicitor; a Jungian analyst from Switzerland; a couple of ecologists; a dance instructor from Edinburgh.

"We are going through a great period of narrative breakdown," said Paul Kingsnorth one afternoon. We were sitting in the lodge house around which we'd pitched our tents, apparently one of the most remote buildings in all of Europe—a forty-five-minute SUV drive across nauseatingly bumpy terrain from the perimeter of the wilderness reserve itself. There was a wood fire crackling and hissing in the hearth, and it was deathly quiet out in the darkening valley, and we were, all sixteen of us, in various attitudes of repose on the couches and chairs and on the floor.

Years back, in a previous life as a journalist in London, Paul had been deputy editor of *The Ecologist* magazine. His old boss, he said, had suffered a nervous breakdown on the job, because the news was all so relentless and unthinkable. Everything was another river drying up, another species disappearing from the Earth.

"There is," he said, "a growing sense of panic and confusion. The stories that we believed aren't true anymore, but we don't know what's true instead." There was, in his manner, a kind of quiet excitement. This moment of painful chaos presented an opportunity to find new stories, new ways of living. It was clear that Paul had come to relish the disruptions of our time, took a kind of perverse satisfaction in the overturning of old orders.

Another Englishman, the former business improvement district manager—whose name was Neil, and who spoke and carried himself in a slow and priestly manner that I found unaccountably touching—spoke in his turn. He said: "There is something about this place that unsettles me very deeply. This is a postapocalyptic landscape. It's a site of total ecological collapse."

Among the group, this sentiment was met with general agreement. It was a beautiful place, but its beauty was cold and unyielding, and largely empty of animal life.

"It is," said Paul. "We're on the edge of civilization out here, in a place that's been stripped bare of life by civilization. That's part of the reason we chose it."

By "we" he meant he and Andres Roberts, the wilderness guide with whom he'd arranged this retreat. Andres was yang to Paul's yin: a cheerful man with a soft Liverpool accent and a quiet but potent charisma, and an uncanny knack of shaping and focusing the group's energy by subtle modulations of his own manner—a shift in posture, a mischievous grin, a bowing of the head, gentle and solemn.

"In a way," said Paul, "this place is a Ground Zero of the industrial age. All the trees in these hills were cut down to provide fuel for industry, to build ships for colonial expansion. An entire attitude toward nature and toward the world spread outward from this place we're in, these islands."

Someone else brought up the Great Oxygenation Event, which had happened about two and a half billion years ago, a mass extinction from which all subsequent life on Earth had evolved. Back then, the world was populated exclusively by single-cell organisms, which lived beneath the surface of oceans that were bloodred due to the massive levels of iron in the water. These microbes relied exclusively on anaerobic methods of respiration—until one species, the cyanobacteria or blue-green algae, began to use the Sun's light to generate vastly more energy than its anaerobic colleagues, by which method it thrived and increased its numbers exponentially, creating via the disruptive innovation of photosynthesis an exploding surplus of oxygen in the planet's atmosphere, toxic to almost every other living thing on Earth. This one rogue microbe changed the atmospheric constitution of the Earth, causing the obliteration of most existing life on the planet and preparing the way for the evolution of multicellular organisms such as ourselves.

"We sort of are those bacteria," said Caroline Ross, an artist who resided on a riverboat on the Thames. "What we are living through, and causing, is like the oxygenation catastrophe. We are making the carbon catastrophe."

She spoke in a quiet and measured tone of how, some time back, she was visiting her brother and, after an intense argument on whose subject she did not expand, she had wandered into his backyard, feeling furious and heartbroken, and had found among the rocks there the fossilized remains of a sea urchin, a species that had, she said, been wiped out many millions of years ago, four mass extinctions before our own. It was a beautiful thing, she said, and holding it in her hand she had felt the slow and inexorable relinquishment of her anger and sadness. She thought of that fossil often, she said, and when she did so she wondered whether we humans would ourselves make good fossils, beautiful imprints in the geological record for some unimaginable future species to wonder over, causing it to think about its own passing from the Earth, its own infinitesimal presence in the dizzying vastness of time. She said that sometimes, in her darker moments, she wished that humans would just cease existing already, or dwindle to a hundred thousand or so in number.

"It's all going to come to an end, and that's okay," interjected a woman with a refined accent. She had a cascade of dark hair, fashionable glasses; she lived in London and made films that were more or less experimental in form. "Nature will reemerge from this, and recover, and it will be beautiful. On some level we are a cancer, and the world will cure itself of us. I want to enjoy the life that I have left. I want to sow good seeds."

I couldn't stop thinking about Caroline's question, about whether we would make beautiful fossils. For all its darkness, what had unsettled me in her slow and measured monologue was how it seemed to come from a place not of misanthropy, but of deeply wounded love—for the world, and for people, too, despite the violence they had done to it. And there was something in this contrast, in her gentleness and despair, that drew me in. I myself had, from time to time, been known to turn my mind to the future extinction of our species, and to the many ways in which creation in its entirety might be better off without us.

One evening around that time I took part in a public discussion about the future of humanity—the topic, more or less, of a book I'd recently published—and after the discussion had ended, a damp and pallid young man had cornered me to give me his thoughts on the matter. He said that in five billion years or so, the sun, having burned through all the hydrogen in its fiery core, would be transformed into a red giant and would expand to engulf much of the solar system, likely even burning up the Earth itself in its explosive demise, and so it was—"obviously," as he put it—necessary to put in place a strategy whereby humanity could continue to survive on some planet far from our own doomed world. I told him that it seemed to me a long shot, given the way things were shaping up—by which I was alluding to the comparatively modest self-inflicted temperature rises we were facing in the coming decades—that our species would survive long enough to witness the consumption of the world by literal cosmic fire. But what I wanted to know, I said, was why it was obvious that we needed an exit strategy, that we should want to survive indefinitely as a species. It caused me no real sadness, I told him, to think that humanity might not exist five billion years from now. I found myself, on the contrary, strangely cheerful about the prospect. Couldn't we just view the eventual death of the sun as an opportunity to call it a day, cosmically speaking? The man looked at me with what seemed like profound bafflement, and suggested that the attitude I'd just outlined was deeply ethically unsatisfactory, given all the future humans who would, in such a scenario, never come to live. He couldn't understand, he said, why I would be okay with humanity as a whole ceasing to exist. Did this not, he asked, fly in the face of a humanist philosophy? I had not said anything about being a humanist, and was in fact not sure I would want to describe myself as such, but I let the matter slide. It seemed to me that we were facing each other across a vast philosophical chasm, one that would not be breached in this conversation, or any other we might be likely to have.

Now and then over the succeeding days, as we walked the hills and valleys, Caroline and I fell into step with one another. She was on more intimate terms with nature than anyone I'd previously encountered, and I was impressed by her extraordinary knowledge of trees and plants and, in particular, species of mushroom. She described herself, half-jokingly, as a Womble, in reference to the 1970s BBC children's television show about furry creatures who lived beneath Wimbledon Common, where they hid from human beings, of whom they generally had a low opinion, and turned their refuse into useful items.

"I make good use of the things that I find," she said. "The things that the everyday folks leave behind."

These words, she explained, were taken from the show's theme song.

She had gone to art college, and had practiced as an artist for a time, before succumbing to what sounded like a deep despair about the futility of producing, of putting more objects into the world, which was, she felt, the last thing the world needed. After years as a musician, singing with various London post-rock bands, it was only fairly recently that she'd gotten back into making art, working solely with materials she had made herself, and making those materials only out of things she found in nature, or that had been discarded by other people—pens made from gull feathers, sketchpads made from pulped linen rags and threaded with bark strips, ink made from oak galls.

An oak gall, she explained, was a bulbous protuberance found on the branches of oak trees, caused by secretions of wasp larvae. From the days of the Roman Empire up until the Industrial Revolution, these were the primary source of ink, but in the last year or so they had become more expensive and difficult to source, she said, on account of the online vaginal health community mysteriously deciding that these wasp nests possessed certain potent vagina-tightening properties, leading to them being sold on Etsy for exorbitant sums. In order to make art out of premodern materials, she said, she now had to get her oak galls from a seller in Germany who imported them from Southeast Asia.

"You can't ever return to the pristine place," she said, with a certain rueful humor. Even a small measure of aesthetic abstinence from modernity required submission to its operations.

I had cause to reflect upon this delicate balance of resistance and accommodation when I noted a small animal skin pouch laid on the table in front of her during one of our conversations. When I asked her about it, she told me it was her smartphone case. She had made it, she said, using the materials and techniques that would have been employed by a Neolithic craftsperson, had there been any requirement for smartphone cases in the Neolithic era.

One evening, Caroline told me about how she'd recently become preoccupied to the point of obsession with Easter Island. She was fascinated in particular, she said, with the idea that the demise of the once-thriving island civilization formed an uncanny reflection of our own particular impasse. There was a theory, she said—albeit one that had been fiercely contested by many historians—that the heads themselves, the giant humanoid constructions known as moai for which the island was primarily known, had been a major cause of the civilization's collapse. When the first Polynesian settlers arrived on the island in the thirteenth century, it was a lush and densely forested environment. Over time, though, population growth and environmental degradation caused by agriculture led to fierce competition over resources, and to tribal conflicts. Deforestation was greatly exacerbated, according to this theory, by the relentless construction of the moai. The construction and transport of these gigantic monoliths, built by tribal chiefs in veneration of their ancestors and as symbols of their own prestige, required massive quantities of wood. Even as the evidence of ecological collapse became overwhelming, the islanders kept constructing the monuments, kept chopping down trees in order to transport them, until there were no more trees to chop. By the time the first Europeans arrived in 1722, soil degradation and deforestation had caused a total collapse, and the population of the island was down from its peak of ten thousand to a few hundred.

Caroline was convinced, she said, that what had happened on Easter Island was what was happening right now, what we were doing to ourselves. Our whole planet, she said, was Easter Island. Here we were, she said, doggedly persisting in the practice of our idolatrous consumerism, heedlessly continuing in the way of life we knew to be causing total ecological collapse, knowing full well the gravity of its consequences, persisting until the last tree was gone, until the soil could no longer support life.

"The way we build our gods," she said, "is the way we build the apocalypse."

She was, in her way, a kind of prepper, though she had nothing but contempt for the actual doomsday survivalists she frequently encountered in her involvement with the bushcraft scene. They were always men, she said, and they came along to classes, but they didn't seem to want to learn. They were interested ultimately not in making things but in equipment. They were always talking about their kit, she said, about their stockpiled foods and their secure locations, about their plans and preparations for absolute self-sufficiency should the shit hit the fan. But the fact of the matter, she said, was that if civilization did collapse these men would be entirely useless to themselves, and worse than useless to everyone else. What they didn't understand, she said, was that the thing that would allow people to survive was the same thing that had always allowed people to survive: community. It was only in learning to help people, she said, in becoming indispensable to one's fellow human beings, that you would survive the collapse of civilization.

She knew what every plant was, every fungus, and took a quiet pleasure in informing you whether it was edible or whether it would kill you. She could probably survive alone in nature if she needed to, she said, and I believed her. But this didn't mean, of course, that she would necessarily want to. She told me one evening about a little carved wooden box she kept locked away on her boat on the Thames, in which she stored thirty seeds she'd gathered from a yew tree. A handful of these, she said, would cause almost immediate heart failure and death. They were an insurance policy against the worst that might happen.

I'd had a weird experience at Heathrow Airport about a month before that trip to Alladale. My first book had been published earlier that year, and I'd been flying a lot in the period since, to book festivals and other events, and running beneath the white noise of my days was a low hum of guilt and shame about the damage my own individual life was doing to the world, the future. In accordance with my anxious custom I had arrived at the airport much too early, established myself at the conveyor belt of a Yo! Sushi. I drank a Japanese wheat ale, and then another. I racked up a tottering stack of color-coded dishes, consuming marine species in delicate arrangements, mackerel, salmon, crab, octopus, tuna. Everything was in season; everything presented itself for immediate acquisition from the moving platform that snaked around the bar, renewing its lavish stock as if by some fairy-tale mechanism of self-replenishing bounty. I was aware of the rapidity with which people were coming and going, racking up their own little stacks of dishes, before grabbing their briefcases, their flight bags, their backpacks, and hurrying for the gates. I had been sitting there for perhaps an hour, longer than the intended Yo! Sushi dining experience, when I became aware that my heart was racing, that I was experiencing a kind of abstract terror. I looked out over the heaving open-plan restaurant, with its ill-defined borders against other similarly heaving restaurants, a gastropub "experience," a Heston Blumenthal–branded solution for on-the-go molecular gastronomy, a high-end meat-and-two-veg concern. In that delirium of commerce, the whole thing lay revealed to me in all its efficiency and excess, its bleak luminescence. I looked at the bright little plates of fish and rice and seaweed and meat as they sailed across my field of vision, cruising their way smoothly around the room to be plucked deftly now and then off the conveyor belt by mostly lone travelers, doughy and exhausted men in pinstripes, young couples in loose-fitting leisure attire, and I thought about the volume of animal and human flesh that was required to keep this system going, the raw tonnage of fuel needed to extract the fish from the sea and ship them to where they were processed, to get them to the gaping mouths of my fellow consumers. All these humans moving between all these places, all this ceaseless motion and consumption, all this hunger and money and flux: it was miraculous and terrible. And it simply couldn't last, was the obvious thing, it simply could not be maintained. The sheer weight and velocity of the system, all of it precariously undergirded by God knew what shifting substructures of finance and power.

An airport is a place in which time and personal autonomy are suspended, in which the only freedom you possess is the freedom to make purchases. The aggressive automation of labor; the nightmare synthesis of fevered consumerism and authoritarian surveillance; the apocalyptic frisson of knowing that all this exists in service of, and is dependent upon, massive rates of carbon consumption. And always, too, the distant limbic hum of death, the screaming descent of the burning jet, as the situation's presiding possibilities, the Chekhovian pistols unholstered at security and irrevocably introduced into the psychic theater of air travel. The oppressive space of the airport—the *junkspace,* to use Rem Koolhaas's unimprovable term—is the architecture of the future itself.

I kept returning to the Heathrow sushi revelation, internally and in conversations with others, because I encountered it as a sort of wound. I encountered it, I mean, as both a realization of the wrongness of our way of life and a mournful intimation of its future passing.

I mentioned it to Andres one afternoon as we sat cross-legged in the grass, and he said it made him think of a graph he had once seen that illustrated the rate of increase of resource consumption throughout the twentieth century and into our time. In the years after the Second World War, he said, the line of consumption had begun to shoot skyward at a vertiginous rate, and looking at it, he experienced a kind of fearful swooning, as though he were gazing downward into an abyss. Looking at the near-vertical line on a page, he said, he felt that he had come into a direct encounter with the absurdity of our world, our way of life.

This was something I myself struggled with, I said, this sense of total absurdity. I felt disinclined to relinquish hope in the world, even incapable of doing so since becoming a parent, and yet the rational part of my brain, the graph-reading part of my brain, insisted that the future was intolerably dark.

There was a paradox at work, I said, in the uneasy depths of my life. The experience of becoming a parent had illuminated that encroaching darkness, made it appear closer to the margins of my own life, and yet in that time I had felt the unmistakable stirrings of hope for the future. I was aware of the possibility that this was a psychological defense mechanism, a denial of the unavoidably obvious, and yet it was no less potent for that awareness. I wondered, in fact, whether there was not some deep selfishness in operation here, some covert mechanism of human delusion, whereby the very fact of having brought a child into a world on the verge of darkness was what had forced me to have hope. And so maybe my own increased sense of optimism about the future had been acquired at the expense of my son, who would now, having been born, have to live in that future.

Andres then spoke about an idea he'd gotten from the

Vietnamese monk and activist Thich Nhat Hanh, whereby there were three circles of care, widening out concentrically like ripples on the surface of water from a dropped stone. There was, at the center, the small circle of the self, and around that the circle of family and friends, and then around that in turn the wider circle of the world. People who care deeply about the world, he said, activists and so on, have a tendency to get angry and burned out from continually fighting battles. And what was needed, he said, was to return to the smaller circle of family and friends, to invest one's energy there, which would in turn bring more generative energy to one's work, to the interaction with the wider circle of the world.

One mercifully mild midday toward the end of the week, each of us set out on our particular paths into the hills and valleys, alone with our tents and our backpacks. This divergence was, in a sense, the whole point of our coming together in the first place, the idea being that we would all strike out into the wilderness in search of a place from which we could see no other humans or signs thereof, there to set up camp for fully twenty-four hours. There would be no distractions: no books, no phones, no conversations or other interventions between ourselves and our situations. I myself went down into the valley and walked along the river for perhaps thirty or forty minutes until I found a little hillock by the bank of the river, flat enough and broad enough to accommodate my tent. Others were more adventurous, heading up the rocky paths and slopes toward the peaks of mountains, toward lakeshores and craggy perches, but I sought the comfort and navigational surety of the water, on the principle that you know where you are with a river, unless you are in it.

I raised my tent as soon as I found my spot, fearing that if I left it until I felt like turning in I might, in the absence of a capable Australian neighbor, come to grief in the cold and dark of a descending highland night. Having thus established myself, I collected some rocks and stones and marked out around myself a circle of roughly ten meters in diameter, within which I committed to remain for the next twenty-four hours. This was the central principle of the ritual: you found a place, and you stayed in it, and you did nothing while you were there. It was known as a "nature solo," and it was inspired by the practice, common to a great many cultures throughout history, whereby an individual went out alone into the wilderness for a time, in search of insight or wisdom or peace.

There was the Vision Quest, the Native American rite of passage in which young men on the cusp of adulthood were sent out by their elders to commune with the spirits of a place, and to seek by way of a vision some part of their ancient wisdom. Andres had spoken of an Icelandic tradition of "going under the hide," whereby a person went out into the wild to "just have a bit of a think about things," bringing an animal skin with them for shelter. ("Someone would ask, where's this or that member of the community gone, I haven't seen him in a bit. And people would say, 'Oh, him? He's gone under the hide.' ") We were more or less unique in our own time and culture, in fact, in having no such common ritual. It was Andres's contention that the wilderness solo was not just a transformative experience for an individual, a means of reaching a deeper connection with the wild and with oneself, but a practice that, were it to be widely adopted, would change our culture's entire relationship to nature.

"The way we live in our everyday lives," he told me one day, "is in disenchanted form. Increasingly since the Scientific Revolution, since Descartes and Newton, we have treated the world as a kind of machine that we control and manipulate, that we change with mechanical movements and levers."

This was our problem as a civilization, in his view, or the root of it. And this was what the solo was about. How were we supposed to care for something if we didn't know it? If we were not, as he put it, in any kind of personal relationship with it?

Having marked out my circle, I sat down on the grass and became immediately preoccupied by the question of how I was going to pass the next twenty-four hours. The view that my situation presented me with, the slope of a mountain descending sheer and grassy toward the river, was undoubtedly very beautiful, but I really couldn't see it holding my attention in the long term. As regards activity, my options were extremely limited. I understood, of course, that having nothing to do was a significant dimension of the solo experience, that it was in a way the whole point of it, but now that I was in the situation, as opposed to merely thinking about it, it had come to seem radically untenable. I am, in theory, a huge fan of doing nothing. As an option, I will typically go with it in any situation where I'm supposed to be doing pretty much anything. But my method of doing nothing, I reflected as I reclined on the soft grass, in fact almost always involved the doing of something, however meaningless or untaxing—scrolling through Instagram or Twitter on my phone, drinking coffee, reading a book or a magazine, going for a walk. None of these things were now possible. I did have my phone with me, but I hadn't had any mobile coverage since we'd left Inverness on Monday morning, and its only practical use at this point was as an alarm clock to alert me, at noon the following day, to the fact that the solo was over. I'd even deleted the *New Yorker* app on my phone, for fear that I might be tempted to fire it up and start reading through back issues of the magazine, immersing myself not in nature, but in long-form reportage.

The only possible activity, in fact, presented itself in the form of half a packet of Marks & Spencer nut and berry mix I'd decided at the last minute to bring with me. This decision ran somewhat against the spirit of the enterprise, in that, although there were no rules or guidelines as such, Andres had recommended that people not bring food, for the simple reason that a packed sandwich or Tupperware container of bean salad or a packet of nut and berry mix would, even if you weren't actually eating it—in fact especially if you weren't actually eating it—become a disproportionate focus of mental energy, in that if you'd decided to wait until late in the evening to eat the sandwich or container of bean salad, the whole rest of the day up to that point would become a kind of prelude to the eating of the snack, and you would find yourself being able to think of almost nothing else until you did eat it—not because of hunger per se, but because of a hunger for something, anything, to occupy your time. But it seemed to me that hunger itself would present a far more formidable distraction than the prospect of its alleviation ever would. If I were to decide to go twenty-four hours without eating, I would be completely preoccupied by the physical sensation of hunger, and increasingly consumed with irritation at myself for having created such a situation, or, what would probably be even worse, utterly consumed with smug self-regard about this act of ascetic self-sacrifice. And this was why, I reasoned, I should eat my half packet of Marks & Spencer nut and berry mix as quickly as possible, in order to get it out of the way and prevent myself obsessing a moment longer than was absolutely necessary.

And so it was that about ninety minutes into my solo, and perhaps an hour after having set up my tent, I established myself in a comfortable sitting position and began to eat, slowly and with uncharacteristic focus, my half packet of Marks & Spencer nut and berry mix. I found myself relishing in particular the cranberries, which I chewed in a very deliberate and controlled manner, and which seemed to me sweeter and more succulent than I had ever remembered cranberries tasting. It occurred to me then that, despite the popular perception of hunger as the best sauce, boredom was perhaps an even better sauce. I soon began to worry, however, about whether I was getting too much pleasure from the Marks & Spencer nut and berry mix—whether I was drawing the snack out to a ludicrous degree in order to forestall the moment when I would have to properly engage with my situation, namely the total immersion in nature. And so I quickly scoffed the remainder of the mix, chiding myself for using the snack as a means of evading the reality of my situation.

The reality of my situation, to be clear, was that I was stuck there, with nothing to do, for a full day and night, in what was probably the most remote location in the entire British Isles, and so there was nothing for it but to begin in earnest my immersive experience of nature. The problem was that I had no idea how to go about having this experience—whether it was something that would simply happen, as it were naturally, and as a direct consequence of my simply being present in nature, or whether some kind of action was required on my part, the deliberate cultivation of an inner state of openness and receptivity. Perhaps, I thought, the two things were not mutually exclusive. I took off my hiking boots and my alpaca wool hiking socks and walked barefoot around the inner perimeter of my little circle a couple of times, focusing all my attention on the sensation of the grass beneath my feet, which was cool and damp, and not entirely unpleasant, but not exactly outright pleasant either. Andres had at some point mentioned that, while doing Qigong exercises out in nature, he always removed his shoes and socks, because it gave him a sensation of "rootedness" in the place he was in, in the Earth itself. This idea appealed to me in theory, but in practice I found that in my bare feet I was unnecessarily preoccupied by the possibility that I might step on a jagged piece of rock, or God forbid an ants' nest, and so in a gesture of compromise I put my alpaca socks back on, though not my boots, reasoning that the socks were at least made from entirely natural materials and so would constitute at worst a kind of buffer zone between myself and nature. I then sat down in front of my tent, assumed the lotus position, and passed perhaps a further twenty minutes to half an hour in failing miserably to focus on my breathing.

At one point, I looked down and saw a tiny creature crawling up the length of my forearm. I had no idea what it was, this creature, though for once in my life I felt no immediate inclination to rid myself of an insect's company. I observed its halting progress toward the crook of my elbow, wondering vaguely what its intentions were, if any, until it suddenly struck me that the creature might well be a tick, and I flicked it off with my forefinger, instinctively rubbing the recently vacated patch of forearm with the palm of my hand. There had throughout the week been a certain amount of low-level hysteria about ticks. We'd been advised to check ourselves for their presence first thing in the morning and last thing at night—because there was a large number of deer in the area, and where there were deer there were probably ticks, and where there were ticks there was the possibility of Lyme disease—cases of which I had recently read had exploded since the 1990s as a result of climate change. Before traveling, I'd done some research about ticks and had watched a video that explained how they operated. They were pretty amazing animals, in a lot of ways. They sense the presence of humans and other large animals by the carbon dioxide we give off. Once they've alighted on our skin, they crawl around in search of a suitable location on which to break the surface and begin feeding. Unlike mosquitoes—whose custom it is to land, get stuck in, and get out of there within a matter of seconds—ticks take their time. It can take them an hour or two to pick the right spot, like tourists with too much time on their hands who can't decide on where they want to eat. Most ticks, I learned, live about three years and eat only three times in their lives—one for each stage of development, from larvae, to nymph, to full adulthood—a fact that, it seemed to me, completely vindicated their seemingly excessive fastidiousness. Once it settles on the right spot, it gets out its elaborate eating gear, including two sets of hooked proboscis, with which it digs into the host's skin, pushing the flesh out of the way and holding it aside to allow for the entry of the harpoon-like hypostome, anchoring the tick firmly in the flesh and allowing it to extract its feed of blood, which blood it prevents from clotting by excreting into the host its own home-brew anticoagulant. Provided the tick isn't found, it will often stay there glutting itself, and swelling to comparatively gigantic proportions, for up to three days, at which point it simply rolls off and goes about its business.

Though I was by no means keen to play host to such a creature—and even less keen to accommodate Lyme disease, with its fevers and its facial paralysis and its debilitating agonies—I could not help but feel some sympathy and respect for its methods. It behooves us humans, I've always felt, to grant at least some grudging admiration to the humbler parasites, on the basic game-recognize-game principle. Their attitude toward us, after all, is strikingly similar to our own approach toward the world in general. Consider the mosquito, statistically the only animal more deadly to humans than we are to ourselves, causing almost twice as many deaths per year as are caused globally by homicide. Mosquitoes have no more against us than we do against the countless species whose extinctions we have caused through hunting or habitat destruction. We simply have something they need in order to live: blood. And the means by which they extract it from us is, it seems to me, uncanny in its similarity to the way in which we ourselves extract minerals from the Earth. To watch a close-up video of a mosquito biting a human—separating its proboscis out into a mechanism of serrated needles, some of which it uses to make deeper incisions into the flesh, others to hold the flesh back for ease of extraction—is to witness something weirdly reminiscent of a sophisticated mining operation. Mosquitoes and ticks and other bloodsucking insects, I thought, checking myself uneasily for any further creatures that might have designs on the precious nectar beneath my skin, are our dark doubles, our brothers in ingenuity and destruction.

What felt like at least an hour must have passed—although it may admittedly have been no more than fifteen or twenty minutes—wherein my sole occupation was the fondling of grass. I ran my fingers through it, plucked it and scattered it to the breeze, held up individual stalks to the light of the sun and inspected them at frankly absurd length. I was deriving a certain aesthetic pleasure from this activity, and had even begun to feel as though I might well have entered into a state of mindful receptivity. I seemed to myself suddenly like a character in a film by Terrence Malick, luxuriating at length in the unconsidered minutiae of nature, cultivating in a moment of complete stillness a kind of quiet aesthetic rapture that verged on the spiritual. But then, of course, it occurred to me that a character in a film by Terrence Malick would never entertain this kind of notion of himself, would never think of himself as a character in a film. What I was, I thought, was a character in a television advertisement whose director was shamelessly, and perhaps even to the point of plagiarism, influenced by Malick, which is to say that my experience was a cheap imitation of the kind of authentically intimate experience of nature you might see in a film by Malick, a filmmaker for whose work I had never had much time to begin with.

As the afternoon wore on, though, this self-consciousness gradually receded, and I began to be able to look at things—the rippling of the grass in the breeze, the glistening of sunlight on the river—without the fact of my looking at them constantly presenting itself to me as evidence of my communion with nature. For several minutes, I watched a minuscule spider meander haltingly across a page of my notebook, before eventually insinuating itself into the little foldable paper pocket inside its rear cover, so that I had to fish it out by means of the concertinaed cream-colored leaflet inside the pocket, which leaflet I then instinctively and reflexively set about reading. I read about how this notebook of mine was heir and successor to the legendary stationery favored by such giants of nineteenth- and twentieth-century culture as Pablo Picasso and Ernest Hemingway and Bruce Chatwin. According to the blurb, it was Chatwin, a particularly obsessive fan of this small black notebook, who gave it the name Moleskine. Chatwin's name was firmly associated in my mind with a kind of stylish writerly ruggedness, and I began to wonder whether, just as my notebook was heir and successor to the legendary stationery favored by Chatwin, I myself, given the kind of literary exercise I was currently engaged in, might one day come to be seen as an heir and successor to Chatwin, of whose writing I had admittedly never read so much as a word, but whom I imagined, perhaps even correctly, to be the kind of writer who went out into the wilderness alone, wearing stylishly practical apparel, featuring many pockets for notebooks and other useful appurtenances, and then came back and wrote about it in prose that was as stylish and practical as his apparel.

At which point, of course, I had been led directly back into the condition of self-consciousness I had briefly transcended, or imagined I had. And what was worse, this return to self-consciousness had been mediated by reading—and reading, of all things, high-end advertising copy—an activity that flew in the face of the whole idea of the solo, which was the removal of all impediments to a deep and authentic experience of nature.

I was then assailed, suddenly and with unexpected force, by a sense of my own ludicrousness. It was abruptly not at all clear to me what I thought I was up to, sitting for hours on end by a riverbank in the remote Scottish wilderness, contemplating nature and my place as a human being within it. In fact, that wouldn't have been so bad; it would have been fine, actually. But what I was doing was attempting to contemplate this stuff and failing miserably. I looked eastward up the valley—although it could have been westward, as I had no real sense of my bearings, either literally or figuratively—toward the mountain Paul Kingsnorth had said he was heading toward for his own solo. I wondered what he was doing up there, up on his literal dark mountain. Certainly, I thought, he wouldn't have been reading through the little leaflet that came with his Moleskine notebook and measuring himself against Bruce

Chatwin, whose work he was almost certainly familiar with. I found it hard to imagine him even having a Moleskine notebook. I imagined him meditating for hours on end. I imagined him whittling a crude wind instrument out of a piece of wood he'd picked up in the grass by his tent. I imagined him not even having a tent, just sleeping out in the open. I imagined him receiving from the ether profound insights about our apocalyptic days, our days of trouble, which by the time he descended from the mountain the following afternoon would have been transfigured into finely crafted stories he would relate to everyone with great narrative skill and conviction. I tried to imagine the sort of insights he might be having, but failed to imagine anything at all, because such insights were, apparently, literally beyond my capacity to imagine.

Eventually, I relinquished the whole notion of profound insights, on the grounds that such interior events could not be deliberately cultivated, even if, and perhaps especially if, you had committed to spending twenty-four hours in a single spot by a river in the Scottish Highlands in pursuit of them. And so I lay back in the grass and looked up at the clouds, having released myself from the obligation to have any particular thoughts about the clouds, or anything else. For a long time, I gazed up at them as they drifted eastward—or possibly westward—above the valley, watching their slow and ceaseless changes of formation, their gatherings and partings, and realizing what a strange spectacle they were, these great shifting fortresses of vapor in the sky, and how rarely I ever focused any serious attention on them, or looked at them for any other reason than judging the likelihood of rain. I watched the shadows cast on the mountain by the migrating clouds, the slow and stately progression of darkness across the landscape, and thought about how everything in the world was moving all the time, how nothing was ever still for even a moment. Each individual blade of grass was trembling minutely in response to some outward force or inner energy. To look closely enough at anything, I thought, was to witness its own particular flux, the pattern of its perpetual mutability.

Slowly but perceptibly, the light receded and the darknesses massed, and I heard the strange machine-like burbling of a bird Dopplering to and fro in the air above me. The air began to chill, and I crawled into my tent and rummaged in my bag for my fleece, and when I emerged again I noted the motionless presence of a stag on the shoulder of the mountain. I looked at him a long time, until it was almost completely dark, and he seemed to be standing not just perfectly still but attentively, too, as though he was waiting for some signal to proceed to his next location. He was much too far away to tell whether he was looking at me, though I assumed that he was, on the perhaps somewhat arrogant assumption that I must have presented the most interesting spectacle in his panoramic vista from the top of the mountain. I wondered what opinion of me he might have formed up there on his crag, before realizing that deer probably did not have opinions about things, and that they were much the better for it.

Something had happened here, I thought, or was in the process of happening. I felt differently about this place. Perhaps the thing that was happening was that I had gone slightly mad from boredom. But I was nevertheless aware of a new feeling, a sensation of tenderness toward the place. And what was more, and stranger, was that I felt that this tenderness was somehow reciprocated. It was just as Andres had said, I realized: something about the experience of being alone here, with nothing to do but be, and sit, and watch, and listen, had caused me to feel as though I were in some kind of relationship with the place. I felt seen by the mountain, known and accepted by the landscape. But even the word *landscape* felt wrong to me. It was a visual term, after all, reflecting the way in which we imposed our aesthetic categories on nature, reducing it to a view, a scene. And it occurred to me that I had never really encountered nature as anything other than a landscape—even, and in fact especially, when I was most struck by its beauty, its weirdness, its otherness. Nature was something I stopped the car to get out and have a look at, before getting back in the car and continuing on. It was something that I consumed, experienced, like a cultural product. But this was not what was going on here. What was happening was not entirely, or even primarily, an aesthetic experience on my part. I was not simply appreciating the view, in other words; nor had I been doing so for hours at this point. I was not having a solitary moment and taking in the air, the landscape. In fact I was—in some strange sense that should by rights have seemed creepy—not even properly alone.

In the dwindling light the mountain had come to appear much closer than it was, and for a moment it seemed to me a living thing—not a mound of cold and insensate rock, but an immense animal that had laid its vast bulk across the land and was peacefully asleep, so that I could almost imagine reaching out and placing a hand on its flank, feeling the blood-warmth beneath the soft skin of the land, the dense and heaving muscle, the quiet aliveness of the earth itself. I wanted to curl up beside the mountain as though it were an old dog, to lay my arm across its back, and press my face softly into its side, and sleep.

I awoke early the next morning, having slept more soundly than I'd expected. I considered packing up my tent, as much for something to do as anything else, but quickly decided against it, figuring I had a further five hours or so to go until the solo ended at midday, and not wanting to risk getting stuck out in the rain without shelter. The day was clear and warm now, but it was the highlands in spring, and you never knew what was in store from one hour of the day to the next. I sat and gazed at the mountain, which seemed farther away now than it had yesterday evening. The feeling of strange intimacy had receded somewhat, though I was aware of a residual affinity. It had been real, this feeling. It could not have been otherwise, in the sense that it was hardly possible to imagine you were feeling something you were not. I struggled to connect it with anything I had experienced before. The one thing it seemed related to, I thought, was a sense I had sometimes had in childhood, at the height of a fever, that the world was pressing in upon me, and that every sound I heard was an act of direct communication: that the closing of a door elsewhere in the house, the creaking of a floorboard, the flushing of a toilet, the distant sound of my mother and father talking—though not the words themselves, which were entirely irrelevant—were speaking directly to me, imparting an urgent and insistent message, but in a code that was completely beyond my ability to decipher. The experience was dreamlike, but at the same time almost unbearably intense, so that I feared I might dissolve entirely under the pressure of this onslaught of encrypted significance. Everything suddenly mattered absolutely, resonated with the urgency of its own being. It was an experience of equal terror and exhilaration.

This thing I had experienced the previous evening was different—less intense and urgent, for one thing, less literally feverish—but it had seemed to arise from a similar place, and to suggest a rupture of the boundary between myself and the world. Was this, I wondered, what Freud was referring to when he talked, in *Civilization and Its Discontents,* about what he called the "oceanic feeling"—the sense of the eternal, the limitless, the boundless? Freud could find no evidence of any such feeling in himself, nor the capacity to experience it, but this didn't stop him from expanding on it at considerable length, based on the descriptions of his friend the French writer Romain Rolland. Rolland had told him that he himself was constantly experiencing this feeling, which was, Freud wrote, "a purely subjective fact, not an article of faith; no assurance of personal immortality attached to it, but it was the source of the religious energy that was seized upon by the various churches and religious systems, directed into particular channels and certainly consumed by them. On the basis of this oceanic feeling alone one was entitled to call oneself religious, even if one rejected every belief and every illusion." Freud himself did not agree with Rolland that this oceanic feeling was in fact the source of all religious sentiment. I don't suppose I would describe my experience as religious, either, or even spiritual. There didn't seem to be anything magical about it, so much as it was a kind of insight into the aliveness of the world.

I had practiced meditation on and off over the last few years, and, while there had been frankly diminishing returns with respect to general mindfulness and well-being, it did occasionally have the effect of creating a mood of immersion in the sensory experience of sound, relieving me for a time of my relentless interiority, and cultivating something like Freud's feeling of the oceanic. And so, sitting in front of my tent, I had closed my eyes and slowed my breathing and was striving to focus on the sensation of air filling and evacuating my lungs, and was attuning myself to the heterogeneous sounds of the wilderness—the distant elated whoop of a lapwing, the tinny whine of a passing mosquito, the endless whispered self-assertions of the river—when out of nowhere the world was murdered, obliterated by a great rupture in the sky. It was the loudest sound I had ever encountered, though I felt it more than heard it: an actual physical force, a violence from the heavens. I opened my eyes, and coming low over the river—three hundred feet, two hundred, a hundred—was a jet shrieking toward me at atrocious speed, and in my mind was the possibility of only one outcome, which was immediate and absolute annihilation. I saw the solitary figure of the pilot in the cockpit, the blank visor of the helmet, and I knew that I'd been seen, and I heard myself howl more in exhilaration than in terror, and then the monstrous visitation was gone, swooping skyward from the water, up and out of the valley and away, leaving only the throbbing echo of the wound it had inflicted on the air. I had dreamt this scene many times, I realized, or something like it, the screaming descent of a plane into a city or a canyon or a body of water, but in those dreams the jet was always a commercial liner, and I was always inside it, frozen in terror, alone, watching the ground rise up to meet me, the vast actuality of onrushing death. I was on my feet now, looking at the sky, and I felt as though I had been on the outside of my own recurring dream, and I was laughing uncontrollably, and my hands were trembling, and I felt utterly alive, and almost physically overcome by an elation of gratitude, though I had no earthly sense of whom or what I might be grateful to.

The irony took some time to settle in, but when it did it settled in deep, and I could think of nothing else. Here I was, the farthest into the wilderness I had ever been, in pursuit of some half-conceived notion of the sublime, of an encounter with the stillness of nature, only to be confronted with the apocalyptic force of the machinery of war. It felt like a mysterious and at the same time almost laughably overdetermined epiphany, a sudden obliteration of one kind of truth by another. (Attempting to follow in the footsteps of Emerson, I had come face-to-face with Pynchon.) I had encountered the sublime after all, but in a form entirely other than what I'd hoped for: this was the military-industrial sublime, the divine violence of technology.

This machine that had flown over my head, close enough that it had tousled my hair like a fond uncle, was, I later learned, a Typhoon bomber from the nearby base of Lossiemouth on the North Sea coast, from where, at that time, the RAF flew jets out to Cyprus for bombing missions in Syria.

It was strange, surreally instructive, to have my little retreat disrupted in this way. I had formed a sacred circle of stones around myself, to make a place of stillness and contemplation and communion with nature, and what had been revealed to me was politics in its rawest form. This wilderness reserve, this place ostensibly dedicated to the undoing of human damage, was also a training arena for war. There is no place where you are outside of power.

In that moment the idea of the apocalypse came into sudden and violent focus. It was already the end of the world for the people that fighter jet was likely headed toward. They were experiencing all the things by which I, in my remote and abstract fashion, was preoccupied: the fragility of political orders, the collapse of civilization. Five million of them, fleeing the terror and chaos of their ruined country, meeting the cruel machinery of Europe and its borders. It was always the end of the world for someone, somewhere.

# 7

# THE FINAL RESTING PLACE OF THE FUTURE

Because I wanted to know what the end of the world might look like, I wanted to go to the Zone. I wanted to haunt its ruins, and be haunted by them. I wanted to see what could not otherwise be seen, to inspect the remains of the human era. The Zone presented this prospect in a manner more clear and stark than any other place I was aware of. It seemed to me that to travel there would be to see the end of the world from the vantage point of its aftermath.

I wanted to go, but not alone. A couple of months after the retreat in Alladale, I called my friend Dylan, who lived in London, and who of all my friends was the one I felt would most likely agree to accompany me to Ukraine on short notice. He was his own boss, for one thing, and he was not short of money, and he was also in the midst of a divorce, amicable but nonetheless complex in its practicalities. It would, I said, be a kind of anti–stag party: his marriage was ending, and I was dragging him to the Chernobyl Exclusion Zone for a weekend. As soon as I'd made it, I felt some discomfort about this joke, with its laddish overtones, as though I were proposing the trip for the sheer lols, or as an exploit in extreme tourism, or, worse still, some kind of stunt journalism enterprise combining elements of both. I was keen to avoid seeing myself in this way.

"I haven't told anyone I'm going," said Dylan over the phone. I was at my hotel at Heathrow.

"Why the secretiveness?" I asked.

"I don't need the hassle," he said. "People thinking I'm weird for wanting to go."

"I know what you mean. There's an ethical queasiness to the whole thing. I have issues with that myself."

"Ethical queasiness? No, I'm talking about radiation. It can't be safe."

"Well," I said, "*safe* might not be the word exactly. But I've done a fair bit of reading about it, and apparently as long as you stay in the designated areas and don't wander into hot spots or whatever, you get exposed to less radiation from a day in the Exclusion Zone than you would from a transatlantic flight."

"I don't know that I necessarily buy that," he said. "What is the source of this factoid?"

"Can't remember."

"Is it the company in Kiev that we're paying to bring us to the Zone?"

"Possibly that is the source of the factoid," I admitted.

"Right," he said. "Excellent."

I realized that I had missed these kinds of exchanges, missed being subjected to Dylan's swift and decisive irony. I'd seen a lot less of him since he'd moved to London four or five years ago and gotten married. Ours was a friendship that made little sense on paper—I was a socialist; he'd been a wealthy entrepreneur since we were in our twenties, having cofounded a tech startup while we were college roommates and sold it to a massive American video games company—and yet it had abided where so many others had fallen into disrepair, or collapsed entirely.

Two days later, not far outside of Kiev, my own trust in the tour company as a guarantor of our safety was badly undermined: it had become clear that our minibus driver and guide, a man in his early forties named Igor, was engaged in a suite of tasks that were not merely beyond the normal remit of minibus driving, but in fact in direct conflict with it. He was holding a clipboard and spreadsheet on top of the steering wheel with his left hand (which he was also using to steer), while in his other hand he held a smartphone, into which he was diligently transferring data from the spreadsheet. The two-hour journey from Kiev to the Zone was, clearly, a period of downtime of which he intended to take advantage in order to get some work squared away before the proper commencement of the tour. As such, he appeared to be distributing his attention in a roughly tripartite pattern—clipboard, road, phone; clipboard, road, phone—looking up from his work every few seconds in order to satisfy himself that things were basically in order on the motorway, before returning his attention to the clipboard.

I happened to be sitting up front with Igor, and with his young colleague Vika, who was training to become a fully accredited guide. Vika was reading on her iPhone a Wikipedia article about nuclear reactors. I considered suggesting to Igor that Vika might be in a position to take on the admin work, which would allow him to commit himself in earnest to the task of driving, but I held my counsel for fear that such a suggestion might seem rude, or even outright sexist. (I can only conclude from this that I would literally rather risk death than risk appearing rude or outright sexist.) I craned around in an effort to make subtly appalled eye contact with Dylan, who was sitting a few rows back alongside a couple of guys in their twenties—an Australian and a Canadian who were traveling around the continent together, apparently impelled by a desire to have sex with a woman from every European nation—but he didn't look up, preoccupied as he was with a flurry of incoming emails. Some long-fugitive deal, I understood, was now on the verge of lucrative fruition.

"Lunch," said Igor, pointing out the side window of the bus. I followed the upward angle of his index finger and saw a series of telephone poles, each of which had a stork nesting atop it. "Lunch," he reiterated, this time to a vague ripple of courteous laughter.

About forty minutes north of Kiev, Igor stuck a USB stick into a console on the dashboard. A screen flickered to life in front of us and began to play a television documentary about the Chernobyl accident. We watched in silence as we progressed from the margins of the city to the countryside. Every so often, Igor demonstrated his familiarity with the documentary by reciting lines of dialogue along with the film. At one point, Mikhail Gorbachev appeared on-screen to deliver a monologue on the terrifying timescale of the accident's aftereffects. His data entry tasks now complete, Igor spoke along in unison with Gorbachev. "How many years will this continue to go on?" he intoned. "Eight hundred years! Yes! Until the second Jesus is born!"

Vika laughed, turning toward me, and I chuckled as though I, too, found this amusing, though I did not.

I was unsure what to make of the tone of all this. Igor and Vika's inscrutable jocularity sat oddly with the task they were charged with: to guide us around the site of the worst ecological catastrophe in history, a source of fathomless human suffering in our own lifetimes. And yet some measure of levity seemed to be required of us.

"Any vegetarians?" Igor had asked as we had climbed aboard the minibus at Independence Square. "If you are vegetarian, we prepare special meal of Chernobyl mushrooms." This had received a muted response, and so Igor clarified that he was joking—a task he would have to repeat many times over the next two days.

After the documentary, the minibus's on-board infotainment programming moved on to an episode of the BBC motoring show *Top Gear,* in which three chortling idiots drove around the Exclusion Zone in family sedans, gazing at clicking dosimeters while ominous electronica played on the soundtrack. There were then some low-budget music videos, all of which featured more or less similar scenes of dour young men—a touchingly earnest British rapper, some kind of American Christian metal outfit—lip-synching against the ruined spectacle of Pripyat.

I wondered what, if anything, the tour company's intention might have been in showing us all this content. Screening the documentary made sense, in that it was straightforwardly informative—the circumstances of the accident, the staggering magnitude of the cleanup operation, the inconceivable timescale of the aftereffects, and so on. But the *Top Gear* scenes and the music videos were much more unsettling to watch, because they laid bare the ease with which the Zone, and in particular the evacuated city of Pripyat, could be used, in fact exploited, as the setting for a kind of perverse adventurism, as a deep source of dramatic, and at the same time entirely generic, apocalyptic imagery.

My feelings on all this were already transitioning from discomfort to outright disdain, when the screen began showing a trailer for something called *Chernobyl Diaries*, a horror flick about a bunch of American twentysomethings who are traveling around Europe when one of them starts pressing the case—"You guys ever heard of Chernobyl? You heard of extreme tourism?"—for a day trip to Pripyat, where they are duly menaced, and lavishly murdered, by some apparently supernatural manifestation of the nuclear disaster.

I was being confronted, I realized, with a cartoonish avatar of my own disquiet about making this trip in the first place; these artifacts of apocalyptiana were on a continuum with the project I myself was undertaking. Was I any less ethically compromised because I had come in search of poetic imagery, or of sociocultural insight? Did the literary form of my intentions make the content any less exploitative? And did my determination to directly confront, on the page, exactly these questions make me, in the end, not less but more ethically culpable, in the sense that I was exploiting my own self-consciousness about exploitation for literary ends?

The minibus slowed as we approached the checkpoint marking the outer perimeter of the Zone. Two uniformed men emerged from a small building, languidly smoking, emanating the peculiar lassitude of armed border guards. Igor reached out and plucked the microphone from its nook in the dashboard.

"Dear comrades," he said. "We are now approaching the Zone. Please hand over passports for inspection."

You feel immediately the force of the contradiction. You feel, contradictorily, both drawn in and repelled by this force. Everything you have learned tells you that this is an afflicted place, a place that is hostile and dangerous to life. And yet the dosimeter, which Igor held up for inspection as we stood by the bus on the far side of the border, displayed a level of radiation lower than the one recorded earlier that morning outside McDonald's in Kiev. Apart from some hot spots, which must be known in order to be avoided, much of the Zone is relatively low in radiation. The outer part of the 30k Zone—the thirty-kilometer radius of abandoned land around the reactor itself—is for the most part perfectly amenable to life.

"Possible to use this part of Zone again humans today," said Igor.

Someone asked why, in that case, it wasn't used.

"Ukraine is very big country. Luckily we can spare this land to use as buffer between highly contaminated part of Zone and rest of Ukraine. Belarus not so lucky."

Immediately you are struck by the strange beauty of the place, the unchecked exuberance of nature finally set free of its crowning achievement, its problem child. The road you walk on is cracked with the purposeful pressure of plant stems from below, the heedless insistence of life breaking forth, continuing on. It is midsummer, and the day is hot, but with the sibilant whisper of a cool breeze in the leaves, and butterflies everywhere, superintending the ruins. It is all quite lovely, in its uncanny way: the world, everywhere, protesting its innocence.

"All the fields are slowly turning into forest," Igor said. "The condition of nature is returning to what it was before people. Mooses. Wild boar. Wolves. Rare kinds of horses."

This is the colossal irony of Chernobyl: because it is the site of history's most devastating ecological catastrophe, this region that was once home to 120,000 Soviet citizens has been for decades now basically void of human life; and because it is basically void of human life, it is effectively the largest nature preserve in all of Europe. To enter the Zone, in this sense, is to have one foot in a prelapsarian paradise and the other in a postapocalyptic wasteland.

Not far past the border, we stopped and walked a little way into a wooded area that had once been a village. We paused in a clearing to observe a large skull, a scattered miscellany of bones.

"Moosc," said Igor, prodding the skull gently with the toe of a trainer. "Skull of moose," he added, by way of elaboration.

Vika directed our attention toward a low building with a collapsed roof, a fallen tree trunk partially obscuring its entrance. She swept a hand before her in a stagey flourish. "It is a hot day today," she said. "Who would like to buy an ice cream from me?" She went on to clarify that this had once been a shop, in which it would have been possible to buy ice cream, among other items.

I exchanged a wary glance with Dylan. He was dressed, as ever, for comfort. Black and gray Nike shell suit, box-fresh white sneakers, dark shades: he looked like a Mafia capo who had by way of some implausible comic contrivance found himself touring the Chernobyl Exclusion Zone in a minibus.

Thirty-one years is a long time, of course, but it was still impressive how comprehensively nature had seized control of the place in that time. In these ruins, it was no easier to imagine people standing around in jeans and sneakers eating ice cream than it was, in the blasted avenues of Pompeii, to imagine people in togas eating olives. It was astonishing to behold how quickly we humans became irrelevant to the business of nature.

Igor pointed out the home of a woman he had often taken his tourists to visit. She had returned in 1988, two years after the accident. Like most of the 140 or so permanent residents of the Zone, known to Ukrainians as *samoseli* (self settlers), she was nearing old age by the time of the evacuation and government resettlement and found it difficult to adjust to life outside of the only home she had ever known. It was December of last year, and it was cold, and when he saw no smoke coming from the roof of her cottage, Igor called her name—"Rosalia! Rosalia!"—and heard no reply. He found her dead in her cottage. She was eighty-eight years old, the last remaining resident of her village.

—

Strictly speaking, everything here is tightly controlled. Strictly speaking, visitors are forbidden from entering any of the buildings in the abandoned city of Pripyat—all of which are in variously advanced states of decay and structural peril, many clearly ready to collapse at any moment. Igor and Vika's employer could in theory lose its license to enter the Zone if its guides were caught taking tourists into buildings. It had been known to happen, said Igor, that guides had had their permits revoked. But the company found itself in something of a double bind in this regard, he explained, on account of the proliferation in recent years of rival outfits offering trips to the Zone. If they didn't take customers into the buildings—up the stairways to the rooftops, into the former homes and workplaces and schoolrooms of the citizens of Pripyat—some other tour company would, and what people wanted more than anything in visiting Pripyat was to enter the intimate spaces of an abandoned world.

One of the Swedish men who accounted for about a third of the group's number asked whether any visitors to Pripyat had been seriously injured or killed while exploring the abandoned buildings.

"Not yet," said Igor, a reply more ominous than he may have intended.

He went on to clarify that the fate of the small but thriving tourism business hung in the balance and depended, by general consensus, on the nationality of the first person to be injured or killed on a tour. If a Ukrainian died while exploring one of the buildings, he said, fine, no problem, business as usual. If a European, then the police would have to immediately clamp down on tour guides bringing people into buildings. But the worst-case scenario was, of course, an American getting killed or seriously injured. That, he said, would mean an immediate cessation of the whole enterprise.

"American gets hurt," he said, "no more tours in Zone. Finished."

Tourism to Chernobyl had expanded rapidly over the last decade or so—according to Igor, there were thirty-six thousand visitors in 2016—boosted by popular entertainments using Pripyat as a postapocalyptic verité setting. Films like *Chernobyl Diaries* and *A Good Day to Die Hard,* television shows like the History Channel's *Life After People* (an entire series devoted to the fetishistic representation of nature's reclamation of the built environment after the disappearance of the human species) and video games like *S.T.A.L.K.E.R., Fallout 4,* and *Call of Duty: Modern Warfare.*

This latter game was in fact among the reasons why Dylan was so quick to agree to this trip: it had a certain sentimental resonance for him, as the game his company—a provider of networking software for online multiplayer games—was working on when it was acquired by Activision, the game's developer.

"This is a hugely iconic place in terms of games," said Dylan.

We were gazing up at the same abandoned Ferris wheel we'd seen several times on the minibus that morning—on the *Top Gear* segment, the movie trailer, the music videos. This was the city's most recognizable landmark, its most readily legible symbol of decayed utopia. Our little group wandered around

Pripyat's fairground, taking in the cinematic vista of catastrophe: the Ferris wheel, the becalmed bumper cars overgrown with moss, the swingboats half-decayed by rust.

The park's grand opening, Vika said, had been scheduled for the International Workers' Day celebrations on May 1, 1986, a week after the disaster, and had therefore never actually been used. Beside her, Igor held aloft the dosimeter, explaining that the radiation levels were by and large quite safe, but that certain small areas within the fairground were dangerously high: the moss on the bumper cars, for example, was among the most toxic substances in all of Pripyat, having absorbed and retained more radiation than surrounding surfaces. So moss in general was to be avoided, as were all kinds of fungi, for their spongelike assimilation of radioactive material. Wild dogs and cats, too, presented a potential risk, not because of rabies, but because they roamed freely in parts of the Zone that had never been effectively decontaminated, and carried radioactive particles in their fur.

I leaned against the railings of the bumper car enclosure and then, recalling having read a warning somewhere about the perils of sitting on and leaning against things in the Zone, quickly relocated myself away from the rusting metal. I looked at the others, almost all of whom were engaged in taking photographs of the fairground. The only exception was Dylan, who was on the phone, apparently talking someone through the game plan for a current investment round. I was struck for the first time by the disproportionate maleness of the group: out of a dozen or so tourists, only one was female: a young German woman who was at present assisting her prodigiously pierced boyfriend in operating a drone for purposes of aerial cinematography. (Through the course of two days in the Zone, we crossed paths with three or four other tour groups, each of which was itself a heavily male enterprise.)

There seemed to be a general implicit agreement that nobody would appear in anyone else's shots, due to a mutual interest in the photographic representation of Pripyat as a maximally desolate place, an impression that would inevitably be compromised by the presence of other tourists taking photos in the backgrounds of one's own. On a whim, I opened up Instagram on my phone—the 3G coverage in the Zone had, against all expectation, been so far uniformly excellent—and entered "Pripyat" into the search box, and then scrolled through a cascading plenitude of aesthetically uniform photos of the Ferris wheel, the bumper cars, the swingboats, along with a great many photos employing these as dramatic backgrounds for selfies. A few of these featured goofy expressions and sexy pouts and gang signs and badass sneers, but the majority were appropriately solemn or contemplative in attitude. The message, by and large, seemed to be this: I have been here, and I have felt the melancholy weight of this poisoned place. (#Chernobyl #amazing #melancholy #nucleardisaster)

Pripyat presents the adventurous tourist with a spectacle of abandonment more vivid than anyplace on Earth, a fever-dream of a world gone void. To walk the imposing squares of the planned city, the broad avenues cracked and overgrown with vegetation, is in one sense to wander the ruins of a collapsed utopian project, a vast crumbling monument to an abandoned past. And yet it is also to be thrust forward into an immersive simulation of the future, an image of what will come in our wake. What is most strange about wandering the streets and buildings of this discontinued city is the recognition of the place as an artifact of our own time: it is a vast complex of ruins, like Pompeii or Angkor Wat, but the vision is one of modernity in wretched decay. In wandering the crumbling ruins of the present, you are encountering a world to come. ("Something from the future is peeking out and it's just too big for our minds," says one of the interviewees in *Chernobyl Prayer,* the Belarusian journalist Svetlana Alexievich's oral history of the disaster and its aftermath.)

And this is why the images from my time in Pripyat that cling most insistently to my mind are the fragmented shards of technology, the rotted remnants of our own machine age. In what had once been an electronics store, the soles of our sturdy shoes crunched on the shattered glass of screens, and with our smartphones we captured the disquieting sight of heaped and eviscerated old television sets, of tubes and wires extruded from their gutted shells, and of ancient circuit boards greened with algae. (And surely I cannot have been the only one among us to imagine the smartphone I was holding undergoing its own afterlife of decay and dissolution.) In what had once been a music store, we walked among a chaos of decomposing pianos, variously wrecked and capsized, and here and there someone fingered the yellowed keys, and the notes sounded strange and damp and discordant. All of this was weighted with the sad intimation of the world's inevitable decline, the inbuilt obsolescence of our objects, our culture: the realization that what will survive of us is garbage.

—

"You ever read any J. G. Ballard?" I asked.

"No," said Dylan. "Why, is he any good?"

We were standing beside an empty Olympic-size swimming pool, staring over the edge into the deep end, the inclined floor of which was caked with dirt, glittering splinters of glass and paint, a damp mulch of leaves. An illegible graffiti throw-up, bubble-style, extended across the near width of the pool.

"He's all right," I said. "A bit repetitive. But an absolute fiend for the symbolism of drained swimming pools, is the reason I ask. This whole place would have been right in his wheelhouse."

I took a couple of photos with my phone, but realized that whatever images I produced would be identical, or inferior, to dozens of others on Instagram, and consequently stopped bothering. I opened my phone's browser and found a picture of the pool in happier times—not, surprisingly, from before the accident, but from the mid-1990s, when it was still used by the so-called liquidators, the military and civil personnel who in those years were charged with cleaning the abandoned city of toxic waste. The shimmering blue water in the photo was no more, and the glass panes of the front wall were all gone, and the ceiling tiles, too, leaving exposed the metal grid of the building's structure. But I was struck by how essentially similar the place was in its state of ruin to what it had been before. Even the clock still hung on the wall at the far end of the pool. It was a large octagonal-faced clock, more or less identical, I realized, to the one that hung on the wall over the pool I regularly swam in near my house—a clock on which you could read both the time of day and, via a large red continuous second hand, the pace of your own laps of the pool. This particular clock, the one I was looking at now, had stopped, whereas the one at the pool I used was, presumably, still counting its seconds. It was another place, Pripyat, another time, and yet entirely recognizable as our own. It was a vast memento mori, a seventeenth-century Vanitas on the scale of a city, a culture.

Dylan zipped his tracksuit top up swiftly and decisively—a gesture that subtly conveyed that he was just about ready to stop contemplating the apocalyptic resonances of the empty swimming pool and move on to the next thing.

"Well," he said, "it's a lot to take in."

We were in the dank foyer of a high-rise apartment building. One of the Swedes, a guy in his late thirties who worked as a school bus driver and seemed notably less enthusiastic about this jaunt than his friends, was standing by a stairwell gazing down at a pile of broken ceiling tiles. His expression was one of mild and obscurely humorous alarm.

"Asbestos," he said. "This whole place is absolutely full of asbestos. All these buildings."

I wasn't all that sure what the deal was with asbestos. I knew only that it was not good. I said, "That stuff is highly flammable, right?"

"No," said the bus driver. "The exact opposite, in fact. It's a flame retardant material. But if you breathe in the dust from it, there are all these little microscopic fibers that get embedded in your lungs and you can never get rid of them, and you die horribly of lung cancer."

"Oh yeah," I said. "I knew it was something."

Dylan stepped back from the pile of broken tiles and politely petitioned the attention of Igor.

"Should we be worried about asbestos at all, Igor?"

"If you don't breathe it, no problem," said Igor, with a shrug of aggressive intricacy and duration.

"Okay, but are you concerned about maybe breathing it without necessarily meaning to?"

"Me? No. Many Europeans and Americans, yes, they are concerned. They are more concerned about asbestos than radiation." At this apparent absurdity Igor chuckled and shook his head.

"But not you," said Dylan.

"Not me," said Igor, and set off up a stairway that looked in its own right to be a grave hazard to public health. Dylan gazed at him as he went, and shook his head in quiet dismay.

"It'll be fine," I said, with no conviction whatsoever.

Outside in the street, a small wild dog approached us with disarming deference. Vika opened up her handbag and removed a squat pinkish tube, a snack from the lower reaches of the pork-product market, and presented it to the dog, who received it with patience and good grace.

There was a dark flash of movement on the periphery of my

field of vision, a rustle of dry leaves. I turned and saw the business end of a muscular black snake as it emerged from beneath a rusted slide and plunged headlong for the undergrowth.

"Viper," said Igor, nodding in the direction of the fugitive snake. He pronounced it "wiper."

We were standing at the entrance to one of Pripyat's many schools, a large tile-fronted building on the side of which was a beautiful mosaic of an anthropomorphic sun gazing down at a little girl. Dylan was rightly dubious as to the wisdom of entering a building in such an advanced state of dilapidation. Turning to Igor, he remarked that they must have been constructed hastily and poorly in the first place.

"No," Igor replied, briskly brushing an insect off the shoulder of his camouflage jacket. "This is future for all buildings."

I never once saw him smile, but his face at rest seemed expressive of a stern and abstruse Slavic irony, and there was undeniably a faint glow of merriment in his bulging eyes. He said that it was best that we go through the building quickly, because it was in particularly poor shape and might collapse at any moment. Dylan suggested that in that case he might sit this one out, but Igor countered that he would simply not allow it, and in a moment of what struck me as wildly uncharacteristic submissiveness, Dylan shrugged and trooped in with the rest of us. Although Igor didn't offer any explicit rationale for this sudden imposition of authoritarianism, our assumption was that it was about minimizing the risk of people getting separated from the group, wandering around without Geiger counters, and potentially straying into invisible pockets of high radioactivity.

The school's foyer was carpeted with thousands of textbooks and copybooks, a sprawling detritus of the written word. It felt somehow obscene to walk on these pages, but there was no way to avoid it if you wanted to move forward. Every building in Pripyat had long ago been looted by so-called stalkers—people, usually teenagers and young men, who entered the Zone illegally in order to explore and hunt for valuables and souvenirs—and the chaos of strewn objects we were met with inside these places was the result not of the disaster itself, but of its aftermath.

In Pripyat, you were always stepping on something that had once meant something to a person long gone. Igor bent down to pick up a colorfully illustrated storybook from the ground, and flipped through its desiccated pages.

"Propaganda book," he said, with a moue of mild distaste, and dropped it gently again at his feet. "In Soviet Union, everything was propaganda. All the time, propaganda."

He picked up another book, a thin monochrome text, and flipped through some pages, before showing me a section illustrated with a drawing of protesting industrial workers, bent and immiserated beneath the weight of exploitation. "This is lesson of Karl Marx," he said. *"Das Kapital."*

I asked him what he himself remembered of the disaster, and he answered that there was basically nothing to remember. Though he was five years older than me, he said that I would likely have a clearer memory of the accident and its aftermath, because in Soviet Ukraine little information was made public about the scale of the catastrophe. "In Europe? Panic. Huge disaster. In Ukraine? No problem."

Climbing the staircase, whose railings had long since been removed or rotted away, I trailed a hand against a wall to steady myself and felt the splintering paintwork beneath my fingertips. I was six when the disaster happened—young enough, I suppose, to have been protected by my parents from the news and its implications. What did I recall of the time? Weird births, human bodies distorted beyond nature, ballooned skulls, clawed and misshapen limbs: images not of the disaster itself but of its long and desolate and uncanny aftermath. I remembered a feeling of fascinated horror, which was bound up in my mind with communism and democracy and the quarrel I only understood as the struggle between good and evil, and with the idea of nuclear war, and with other catastrophes of the time, too, the sense of a miscarried future. As I continued up the stairs, a memory came to me of a country road late at night, of my mother helping me up onto the hood of our orange Ford Fiesta, directing my attention toward a point of light arcing swiftly across the clear night sky, and of her telling me that it was an American space shuttle called *Challenger,* orbiting the planet. That memory was linked in my mind with a later memory, of watching television news footage of that same shuttle exploding into pure white vapor over the ocean. The vision of the sudden Y-shaped divergence of the contrails, spiraling again toward each other as the exploded remains of the shuttle fell to the sea, a debris of technology and death, eerily striking against the deep blue sky. That moment was for me what the moon landing was for my parents and their generation: an image in which the future itself was fixed.

We rounded the top of the stairs, and as I set off down a

corridor after Igor, I realized that those images of technological disaster—of explosions, mutations—had haunted my childhood, and that I had arrived at the source of a catastrophe much larger than Chernobyl itself, or any of its vague immensity of effects. Panic. Huge disaster. I remembered a line from the French philosopher Paul Virilio—"The invention of the ship was also the invention of the shipwreck"—which seemed to me to encapsulate perfectly the extent to which technological progress embedded within itself the prospect of catastrophe. And it occurred to me that Pripyat was a graveyard of progress, the final resting place of the future.

In a large upstairs classroom, a dozen or so toddler-size chairs were arranged in a circle, and on each was perched a rotting doll or distempered teddy bear. The visual effect was eerie enough, but what was properly unsettling was the realization that this scene had been carefully arranged by a visitor, probably quite recently, precisely in order for it to be photographed. And this went to the heart of what I found so profoundly creepy about the whole enterprise of catastrophe tourism, an enterprise in which I myself was just as implicated as anyone else who was standing here in this former classroom, feeling the warm breeze stirring the air through the empty window frames. What got to me about Pripyat was not its desolation, nor even the ever-present potential for radioactive toxicity, but rather the sense of the place as fitting neatly into a preexisting aesthetic framework, a sense that in merely being here we were partaking of, were in fact in pursuit of, a kind of apocalyptic kitsch. We were, in other words, consuming a product here. On some level I had understood this all along, but what unsettled me was the failure to hide this uncomfortable fact, the failure even to try. I'd paid two hundred quid for this tour, inclusive of meals and accommodation and transport.

I wondered whether Igor and Vika held us in contempt, us western Europeans and Australians and North Americans who had forked over a fee roughly equivalent to Ukraine's average monthly wage in order to be guided around this discontinued world, to feel the transgressive thrill of our own daring in coming here. If it were me in their position, I knew that contempt is exactly what I would have felt. The fact was that I didn't even need to leave my own position in order to hold myself in contempt, or anyone else.

"How often do you come here?" I asked Igor.

"Seven days a week, usually," he said. He had a strange way of avoiding eye contact, of looking not directly at you but at a slight angle, as though you were in fact beside yourself. "Seven days a week, eight years."

"How has that affected you?" I asked.

"I have three children. No mutants."

"I don't mean the radiation so much as just the place. I mean, all this must have an impact," I said, gesturing vaguely toward my own head, indicating matters broadly psychological.

"I don't see my wife," he said. "My family. I get up at five-thirty a.m., they are asleep. I get home late night, already they are asleep again. I am a slave, just like in Soviet Union time. But now I am a slave to money."

I nodded shrewdly, in a sing-it-sister kind of way, though I was not myself especially a slave to money.

"You know Dr. Alban? The rapper? It's like he says in his song 'It's My Life.' "

I must have looked confused, because Igor clarified: "It's my life. This."

Just then, one of the Swedes emerged from a smaller classroom, a lumbering man in middle age, heavily weighted with a large backpack and a great deal of expensive-looking photographic equipment.

"Did you say Dr. Alban?"

"Dr. Alban, yes," agreed Igor.

"Swedish!" said the Swede, with a certain prideful complacency. "From Sweden."

"Really?" said Igor.

"He is a dentist, you know," said the Swede. "In Sweden, he is in fact a dentist."

"Really?" I said. Granted, I knew very little about Dr. Alban, of the details of his biography either before or after his 1992 pan-European smash hit "It's My Life," but this fact of his being a qualified dentist seemed strongly counterintuitive. "Does he still practice?"

"What?" said the Swede, who was no longer chuckling.

"Does he still practice dentistry, I mean?"

The Swede shook his head in bewilderment, looking at me as if I had just said something completely insane, as if it was I who had just invoked the half-forgotten specter of Dr. Alban and started asking unprovoked whether he practiced dentistry.

"I have no idea," he said, and went back into the classroom he'd emerged from, gazing down at his camera settings.

I went with Igor and Vika into another classroom, where we were followed by the wild dog Vika had fed earlier. The dog did a quick circuit of the room, sniffed perfunctorily at a papier-mâché doll, an upturned chair, some torn copybook pages, then settled himself down beside Vika. Igor opened a cupboard and removed a stack of paintings, spread them out on a table flaked with aquamarine paint. The pictures were beautifully childish things, heartbreakingly vivid renderings of butterflies, grinning suns, fish, chickens, dinosaurs, a piglet in a little blue dress. They were expressions of love toward the world, toward nature, made with such obvious joy and care that I felt myself getting emotional looking at them. I could all of a sudden see the children at their desks, their tongues protruding in concentration, their teachers bending over to offer encouragement and praise, and I could smell the paper, the paint, the glue.

I picked up a painting of a dinosaur, and I was surprised by sadness not at the unthinkable dimensions of the catastrophe itself, but at the thought that the child responsible for this picture had never been able to take it home to show his parents, had had to leave it behind just as he had had to leave behind his school, his home, his city, his poisoned world. And I became conscious then of the strangeness of my being here, the wrongness of myself as a figure in this scene: a man from outside, from the postapocalyptic future, holding this simple and beautiful picture in his hand and looking at it as an artifact of a collapsed civilization. This, I now understood, was the deeper contradiction of my presence in the Zone: my discomfort in being here had less to do with the risk of contamination than with the sense of myself as the contaminant.

Czesław Miłosz's poem "A Song on the End of the World" conjures a last day that is just like any other, where nature continues about its business, and where "those who expected signs and archangels' trumps / Do not believe it is happening now." The poem closes with a white-haired old man repeating the following lines as he harvests tomatoes in his field: "There will be no other end of the world, / There will be no other end of the world."

The poem was written in 1944, in occupied Warsaw. Nowhere in it does Miłosz mention or even allude to Auschwitz, where as he wrote, just a couple of hundred miles away the end of the world had been under way for some time. But it's impossible to read the poem now without thinking of that localized apocalypse. The fact that the world is continuing on as always—that the sun is shining, and the bees circling the clover, and the tomatoes ripe in the fields—doesn't mean it hasn't already come to an end.

The image of Miłosz's white-haired old man, his prophet who is not a prophet, came to me when we were taken to meet Ivan Ivanovich Semeniuk, a farmer in his early eighties who was one of the last remaining returnees to the Zone. He was one of two remaining inhabitants of the village of Paryshev, which had once been home to about six hundred people. (The other was a cheerful and exceptionally tiny old lady named

Darya. Darya lived a short walk across the fields with a small brown terrier.) Ivan Ivanovich carried for some reason an ear of corn in his hand the entire time we were with him, and wore a khaki army jacket and loose-fitting lavender running pants, and though he walked with the aid of a cane he seemed in excellent shape for a man of his age, let alone a man of his age who had lived for the last three decades in a zone of nuclear alienation. He told us, via the medium of Vika's halting simultaneous translation, that when he and his family were evacuated six days after the accident, they were informed, like everyone else in what would become the Exclusion Zone, that they would be returning within a few days, and so they took almost nothing with them, a few bags of potatoes. All their animals were taken away and slaughtered, buried in the ground.

(There was no end to the things that were buried by the liquidators in the days and months and years after the accident: they bulldozed schools and houses, buried the rubble of entire towns, and they buried whole forests, sawing up trees and packing the logs in plastic and putting them deep in the ground, and they buried the earth itself, removing the contaminated layer of soil and burying that, and they buried the bodies of the first responders—men exposed to such severe doses of radiation that they died with blisters on their hearts—in lead coffins welded shut to minimize radioactive leak from the corpses.)

In 1987, after about a year and a half living elsewhere in Ukraine, Ivan brought his family back to their village and this home he had built with his own hands from wood and corrugated iron. Returning was not, strictly speaking, legal, but the government tolerated the two thousand people who decided they would rather risk the consequences of returning to the only land they'd ever known than live healthy but miserable lives in the government-provided apartments in inner-city Kiev. Many of those who were resettled after the accident were isolated and shunned by their new urban neighbors, who were wary of contamination through the physical proximity of these Chernobyl people.

After his return, Ivan Ivanovich worked for a few years as a guard at the power plant, and then as a road builder, before retiring to live off the land with his wife, Maria. She had died the previous year, and he now lived alone, though he had a son in Kiev who visited him often. Ivan Ivanovich grew his own vegetables, gathered mushrooms and berries from the forest around his house, kept chickens and a pig, and he burned radioactive wood in his stove to keep warm, and if this life in the Zone had caused him any serious harm, he had failed to notice it. He observed with a kind of rueful satisfaction that he had outlived many evacuees of his age who had declined to return to the Zone.

"You're living your life," says one of the anonymous interviewees in Svetlana Alexievich's *Chernobyl Prayer.*

> An ordinary fellow. A little man. Just like everyone else around you—going to work, coming home from work. On an average salary. Once a year, you go on holiday. You've got a wife, children. A normal sort of guy. And then, just like that, you've turned into a Chernobyl person. A curiosity! Some person that everyone shows interest in, but nobody knows much about. You want

> to be the same as anyone else, but it's no longer possible. You can't do it, there's no going back to the old world. People look at you through different eyes [. . .] In the beginning, we all turned into some kind of rare exhibits. Just the word *Chernobyl* still acts like an alarm. They all turn their heads to look at you. "Oh, from that place!" That's what it felt like in the first days. We lost not just a town but a whole life.

We followed Ivan Ivanovich around his little property in the manner of visiting dignitaries, respectfully taking note of his kitchen garden and his grapevines and the ancient orange Lada ("Soviet Porsche!" announced Vika) that sat rusting in his garage, and which he assured us would still be absolutely roadworthy had he anywhere to go. We passed a low, shack-like construction made of rusting scrap metal and wood, insulated with black plastic sheeting, referred to by Igor as Ivan Ivanovich's "moonshine reactor," the means by which he was able to keep himself in booze. There was much photography, and many comradely selfies were taken with our smiling host, and it was clear to me that we as a group were at least as strange and remarkable, at least as worthy of anthropological consideration, as this elderly farmer and the dwindling postapocalyptic peasantry to which he belonged.

The tour company had put us up in the town of Chernobyl itself, in a place called Hotel 10—a name so blankly utilitarian that it sounded chic. Hotel 10 was in reality no more chic than you would expect a hotel in Chernobyl to be, and arguably even less so. It looked like, and essentially was, a gigantic two-story shipping container. Its exterior walls and roof were corrugated iron. Internally it seemed to be constructed entirely from drywall, and it smelled faintly of creosote throughout, and the long corridor sloped at a nauseating angle on its final descent toward the room Dylan and I were sharing on the ground floor.

The Ukrainian government imposes a strict 8:00 p.m. curfew in the Zone, and so after a dinner of borscht, bread, and unspecified meats, there was nothing to do but drink, and so we drank. We drank an absurdly overpriced local beer called Chernobyl, which the label assured us was brewed outside the Zone, using nonlocal wheat and water, specifically for consumption inside the Zone itself—a business model that Dylan rightly condemned as needlessly self-limiting. (The hotel had run out of all other beers; it was either this stuff or nothing at all, and it was decided that Chernobyl, despite its comparatively exorbitant price point, was clearly preferable to nothing at all.)

We all turned in early that night. Even if we'd wanted to walk the empty streets of the town after dark, we'd have been breaking the law in doing so, and possibly jeopardizing the tour company's license to bring tourists to the Zone. Unable to sleep, I took out the copy of *Chernobyl Prayer* I'd brought with me. As I reached the closing pages, after dozens of monologues about the loss and displacement and terror endured by the people of Chernobyl, I was unsettled to encounter an image of myself. The book's coda was a composite of 2005 newspaper clippings about the news that a Kiev tour company was beginning to offer people the chance to visit the Exclusion Zone.

"You are certainly going to have something to tell your friends when you get back home," I read. "Atomic tourism is in great demand, especially among Westerners. People crave strong new sensations, and these are in short supply in a world so much explored and readily accessible. Life gets boring, and people want a frisson of something eternal . . . Visit the atomic Mecca. Affordable prices."

After nearly three hundred pages of melancholy monologues of loss and displacement and terror, this was a strange and discordant note for the book to end on. It would have been unsettling enough had I not actually been, at that moment, on the exact tour I was reading about. Alexievich's voice was, strictly speaking, not a presence on the page—as with most of her work, she ceded the floor more or less absolutely to her interviewees—but with this final flourish, it was impossible not to detect a note of exhausted authorial irony, even of disgust about this final indignity being visited on the land, a kind of cultural contamination.

I lay awake for some time, trying to attend to the silence, hearing now and then the faint howling of wolves in the lonely distance, remembering having read somewhere that the Zone was now home to the highest concentration of gray wolves in all of Europe. Had I myself, I wondered, come here in search of strong new sensations? Was I looking for a frisson of something eternal, and was that frisson the radiation itself, the incurable poison in the land, or was it the very emptiness of the place? There was, I realized, a sense in which I was encountering the

Zone less as the site of a real catastrophe, a barely conceivable tragedy of the very recent past, than as a vast diorama of an imagined future, a world in which humans had ceased entirely to exist. And it was the abandonment of the place, its very emptiness, that, paradoxically, exerted such a powerful attraction on me, and on people like me.

Among ruins, Pripyat is a special case. It's Venice in reverse: a fully interactive virtual rendering of a world to come. Its uncanniness arises out of the indeterminate status of its strewn disjecta; the smashed televisions, the rotted pianos, the finger-paintings of its departed children: these are both trash and artifact. The place is recognizably of our own time, and yet entirely other.

It was built as an exemplary creation of Soviet planning and ingenuity, an ideal place for a highly skilled workforce. Broad avenues lined with evergreen trees, sprawling city squares, modernist high-rise apartment buildings, hotels, places for exercise and entertainment, cultural centers, playgrounds. And all of it was powered by the alchemy of nuclear energy. The people who designed and built Pripyat believed themselves to be designing and building the future. This is a historical irony almost too painful to contemplate.

As strange as it felt to be here, as obviously wrong on some level, I was aware of this activity as having a distinct cultural lineage. The taking of pleasure in ruins, perverse though it may be, has been a popular pursuit of centuries. (The Germans, of course, have a word for it: *Ruinenlust.*) From the late seventeenth century onward, Britain's young elite began to undertake their Grand Tours, their customary post-Oxbridge sojourns among the cultural sites of continental Europe, of which Greek and Roman ruins were a central element. In this way, they were reminded that even the greatest civilizations, the greatest empires, must all eventually fall into ruin.

Meditation upon impermanence became a mainstay of eighteenth- and nineteenth-century art and literature. "The ideas ruins invoke in me are grand," wrote Diderot, in his "Salon of 1767." "Everything comes to nothing, everything perishes, everything passes, only the world remains, only time endures." A fundamental aspect of what he called "the poetics of ruins," he wrote, was the way in which spending time in such places caused one to think about how the spaces of one's own life would inevitably come to languish in just this condition of dilapidation. "Our glance lingers over the debris of a triumphal arch," he wrote, "a portico, a pyramid, a temple, a palace, and we retreat into ourselves; we contemplate the ravages of time, and in our imagination we scatter the rubble of the very buildings in which we live over the ground; in that moment solitude and silence prevail around us, we are the sole survivors of a nation that is no more."

I did experience this myself, though it was subject to a delayed effect. It wasn't until after I returned home from Ukraine that I began to imagine my own house a ruin, to picture as I walked through its rooms the effect thirty years of dereliction might wreak on my son's bedroom, imagining his soft toys matted and splayed to the elements, the bare frame of his bed collapsed in a moldering heap, the floorboards stripped and rotted. I would walk out our front door and imagine our street deserted, the empty window frames of the houses and shops, trees sprouting through the cracked sidewalks, the road itself overgrown with grass. For some reason it had not occurred to me in Pripyat that the house I myself lived in was considerably older than any of the buildings I encountered there, older than the Soviet Union itself. And that the foundations of Pripyat's ruins, in fact, were laid barely a decade before I was born.

At the center of the Zone is Reactor Number 4. You don't see it. Not now that it is enclosed in the immense dome of steel and concrete known as the New Safe Confinement. This, they say, is the largest movable object on the planet: 360 feet tall at its apex, 886 feet wide. The dome was the result of a vast engineering project involving twenty-seven countries. The construction had been completed onsite, and the previous winter the finished dome had been slid into position on rails, over the original shelter, which it now entirely contained. That original shelter, known variously as the "Sarcophagus" and the "Shelter Object," had been hastily constructed over the ruins of the reactor building in the immediate aftermath of the disaster, but had in the intervening years suffered corrosion from the elements and had begun to leak radiation into the soil beneath the plant.

The group stood looking at the dome, attending to Igor as he talked us drily through the stats, taking photos of the plant for later Instagram sharing.

"*Sarcophagus* is an interesting word to have gone with," said Dylan, trousering his phone.

"It really is," I said. "They have not shied away from the sinister as regards terminology."

*Zone. Stalker. Shelter Object. Sarcophagus.* There was an archetypal charge to these terms, a resonance of the uncanny on the surfaces of the words themselves. *Sarcophagus,* from the Greek. *Sark* meaning flesh; *phagus* meaning to eat.

A couple of hundred yards from us—beneath the ruins of the reactor building, beneath the sarcophagus, beneath the great silver dome that enclosed it all—was an accretion of fissile material that had burnt through the concrete floor of the reactor building to the basement beneath, cooled and hardened into a monstrous mass they called the Elephant's Foot. This was the holy of holies, the most toxic object on the planet. This was the center of the Zone itself. To be in its presence even briefly was to relinquish your life. Thirty seconds would bring about dizziness and nausea. Two minutes, your very cells would begin to hemorrhage. Four minutes: vomiting, diarrhea, a fever in the blood. Five minutes in its presence, and you would be dead within two days. Concealed though it was, its unseen presence emanated a shimmer of the numinous. It was the nightmare consequence of technology itself, the invention of the shipwreck.

In the closing stretch of the Bible, in the Revelation, appear these lines: "And the third angel sounded, and there fell a great star from heaven, burning as it were a lamp, and it fell upon the third part of the rivers, and upon the fountains of waters. And the name of the star is called Wormwood: and the third part of the waters became wormwood; and many men died of the waters, because they were made bitter." Wormwood is a woody, bitter-tasting shrub that is used throughout the Bible to mean a curse, the wrath of a vengeful God. In Ukrainian, and in other Slavonic languages, the word for wormwood is *chernobyl.* (The plant grows in lavish abundance along the banks of the River Pripyat.)

This matter of linguistic curiosity is frequently raised in commentaries on the accident, its apocalyptic resonances. In one of the long monologues recorded by Alexievich in *Chernobyl Prayer,* the speaker quotes the lines from the Revelation and then says this: "I'm trying to fathom that prophecy. Everything has been predicted, it's all written in the holy books, but we don't know how to read."

Laborers in construction hats ambled in and out of the plant. It was lunchtime. The cleanup was ongoing. This was a place of work, an ordinary place. But it was a kind of holy place, too, a place where all of time had collapsed into a single physical point. The Elephant's Foot would be here always. It would remain here after the death of everything else, an eternal monument to our civilization. After the collapse of every other structure, after every good and beautiful thing had been lost and forgotten, its silent malice would still be throbbing in the ground like a cancer, spreading its bitterness through the risen waters.

Before returning to Kiev, we made a final stop at the Reactor 5 cooling tower, a lofty abyss of concrete that had been nearing completion at the time of the accident and had lain abandoned ever since, both construction site and ruin. Igor and Vika led the way through tall grass, and across a long footbridge whose wooden slats had rotted away so completely in places that we had to cling to railings and tiptoe along rusted metal sidings.

"Welcome to Indiana Jones part of tour," said Igor. Neither the joke itself nor the halfhearted titters it received seemed to give him any pleasure whatsoever. It was a part of the job, like any other: he walked across the rotted footbridge; he delivered the Indiana Jones line; he proceeded to what was next.

Once inside, we wandered the interior, mutely assimilating the immensity of the structure. The tower ascended some five hundred feet into the air, to a vast opening that encircled the sky. In puckish demonstration of the cooling tower's dimensions, Igor selected a rock from the ground and pitched it with impressive accuracy and force at a large iron pipe that ran across the tower's interior, and the clang reverberated in what seemed an endless self-perpetuating loop. Somewhere up in the lofty reaches a crow delivered itself of a cracked screech, and this sound echoed lengthily in its turn.

In the Old Testament, some of God's more memorable threats to various insubordinates, various enemies of his people, involve ruined cities as the terrain of roosting birds. In the Book of Jeremiah, He declares that the city of Hazor, in the wake of its destruction by Babylon, will become "a haunt for jackals, a desolation forever." And then there is the great blood-fevered edict of Isaiah 34, where it is foretold that the Lord's righteous sword will descend on the city of Edom—her streams turned into pitch, her dust to blazing sulfur, her land lying desolate from generation to generation—and that this city, too, will become "a haunt for jackals, a home for owls." They will possess it forever, God says, and dwell there from generation to generation.

The more adventurous of us clambered up the iron beams of the scaffolding in search of more lofty positions from which to photograph the scene. I was not among them. As was my custom, I sought the lower ground, sitting cross-legged in the dirt, having forgotten for a moment the obvious danger of doing so. There was a concrete wall ahead of me, on which was painted a monochrome mural depicting a surgeon in scrubs and mask, hands pressed to his face, eyes staring ahead in deep weariness and horror. This image I recognized as a photograph by Igor Kostin, a press photographer known for his documentation of the disaster and its aftermath. It was an incongruous enough thing, this work of street art inside the abandoned shell of the tower, but it also struck me as trite and banal, as somehow a violation of the integrity of the ruin. It subtracted from the pitiless poetry of the place.

I looked up. Hundreds of feet overhead, two birds were gliding in opposing spirals around the inner circumference of the tower, kestrels I thought, drifting upward on unseen currents toward the vast disk of sky, impossibly deep and blue. I sat there watching them a long time, circling and circling inside the great cone of the tower. I remembered Alladale, the death-thrill of the fighter jet shrieking through the valley toward me, the blank brutality of technology in a wild solitude.

These birds of prey, drifting and mysterious, seemed an equal and opposite revelation, a fleeting disclosure of some hidden code or meaning.

This place is a message. A haunt of jackals, and a desolation forever.

I laughed, thinking of the Yeatsian resonances of the scene, the millenarian mysticism: the tower, the falcons, the widening gyres. But there was in truth nothing apocalyptic about what I was seeing, no blood-dimmed tide. It was an aftermath, a calm restored.

These birds, I thought, could have known nothing about this place. The Zone did not exist for them. Or rather, they knew it intimately and absolutely, but their understanding had nothing in common with ours. This cooling tower, unthinkable monument that it was to the subjugation of nature, was not distinguished from the trees, the mountains, the other lonely structures on the land. There was no division between human and nonhuman for these spiraling ghosts of the sky. There was only nature. Only the world remained, and the things that were in it.

8

# THE REDNESS OF THE MAP

Our daughter was born into a drought. The weather all that year had been strange, volatile, careening between opposing extremes.

Six weeks before her due date, the snow lay deeper over Dublin than it had before in the years of my lifetime, and the airports were shut, and the streets were becalmed, and the shops had run out of sliced bread, and across the city eight men—driven perhaps by panic, but more likely the wild festive euphoria of an extreme weather event—had stolen a digger from a construction site and used it to smash in the back wall of a supermarket in the night. Briefly but memorably, the army was called in. It was funny, in that it confirmed Ireland's sense of itself as a place never far from lawlessness, and comically ill-equipped to deal with anything more extreme than drizzle, but it also suggested how much less funny things might get if the weather stayed weird much longer.

The government issued a Status Red weather warning, advising that people stay indoors unless absolutely necessary. It was a reminder of how fragile everything was, how flimsy the supply chains, how a few days of snow could bring everything to a stop.

Then, in the first weeks of her life, there was the strange heat. Throughout late May and all through June of that year, it was unprecedentedly hot and dry—the most intense heat wave of my lifetime, the longest period without rain. Whenever we brought the baby outside, we worried about whether she was sufficiently shaded from the intensity of the sun. Indoors we struggled to keep her cool. We sat out in our tiny shaded courtyard, with its cast-iron furniture and its potted plants, drinking coffee and iced fruit juice, marveling at this sudden Mediterranean swerve our lives had taken.

My wife reminded me that when we'd bought the house, people had remarked that this outdoor area, which received little direct sunlight, would be a tremendous asset in the heat of southern Europe. Less so in Ireland, they would always laugh. And now here we were, she said, sheltering from the heat like a couple of Sicilians.

Yes, I agreed, here we were. It was funny, and also not at all.

Because as pleasant as it was, this weather, it could not be received as an unmixed blessing. Every day, there were things in the news that gave a person pause. For the second time in three months, the government had issued a Status Red weather warning, this time for the risk of wildfires. Restrictions on water usage were imposed. The tabloids were reveling in a story, obvious publicity gambit though it was, about how the lack of soil moisture would mean a greatly diminished potato crop, and therefore that the country would likely suffer a "crisp shortage" in the coming year. It was a comic recapitulation of Ireland's history of potato blight and famine. First as tragedy, then as farce. Funny, and also not at all.

On the evening news, on the front pages of tabloids, the maps of Europe blazed a vengeful red. The Swedish government was appealing for help from other EU countries to deal with rampant wildfires in the Arctic Circle. On my phone, I watched streaming video of blazing pine forests, of planes and helicopters spreading great billowing spumes of water over the burning land, and I thought about the term "Arctic heat wave," an absurdity that threatened to short-circuit thinking altogether.

That there were wildfires in the Arctic Circle felt like the most important fact in the world. This was a thing we should never not be thinking about, talking about. But something about this truth, and the endless deluge of other more or less equally horrifying truths, made it almost impossible to assimilate. The subtext of every news headline now, of every push notification, was that we were completely and irrevocably fucked.

But Arctic wildfires? This was the subtext erupting through the surface, combusting in the dry heat of overdetermination. This, if anything, was too on the nose.

In Greece, meanwhile, dozens of people had died in wildfires at seaside resorts near Athens. Hundreds had leapt into the ocean to flee the flames, which were whipped into frenzy by winds of sixty miles an hour. Many didn't make it and burned to death on the shore, while many more drowned. The streets were lined with the scorched skeletons of cars, keys still in their ignitions, their owners having abandoned them en masse to escape the approaching inferno on foot.

This was the catastrophe itself, ongoing and absolute, hiding in plain sight. If you asked me how I was, how things were going, I should only have been able to honestly tell you that there were wildfires in the Arctic Circle, because that was really all that could be said about how things were. But humankind, as the bird in T. S. Eliot's "Burnt Norton" famously put it, cannot bear very much reality.

For a few months that year, I shared an office with an ecologist who did consultancy work, speaking to corporations about how they could make their businesses more sustainable. Over lunch one day, I said I was skeptical about the idea that corporate sustainability, and individuals living more consciously and responsibly in terms of their impact on the environment, could at this point have any meaningful impact on what we were headed for.

"I feel like we're fucked," I said. "Are we fucked?"

Though it was not a sentiment she would want to share with her corporate clients, she conceded that we were fucked. The only way she could conceive of our species doing what needed to be done to halt our progress toward catastrophe, she said, was the imminent establishment of some kind of benevolent global dictatorship whose sole purpose was to limit the amount of carbon we released into the atmosphere. This seemed an unlikely prospect, she said.

The most that could be hoped for now, she said, was that

we could find ways to shield ourselves from the worst effects of what was to come. The word she used, I remember, was *sandbagging.*

"When we say 'fucked,'" I said, "are we talking about the same thing? Because I'm talking about the collapse of civilization."

People often asked her, she said, about Ireland's prospects when it came to climate change. What she told them was that we were extremely lucky in a lot of ways, that we were in a very small group of nations—New Zealand being another—that were unlikely to suffer catastrophic effects from melting polar ice caps, a hotter and drier climate. People tended to think, she said, that this meant we would be fine, that we would simply have to become more self-sufficient. But this was pure delusion. What would it even mean, after all, to be fine in the context of a drowning world, a world on fire? We were a small island, with nine hundred miles of coastline and an army that would by itself be effectively useless against any kind of invasion. We would be relying, she said, on the goodwill of other countries whose people were starving, drowning, burning. We would not be fine.

One day in July, *The Sun*'s front page featured a fiery red map of heat waves across Europe and North America, and a headline proclaiming THE WORLD'S ON FIRE. Smaller images displayed the inferno in Greece, the grass in a London park turned brown. (Just above the image of the burning world was a notice of a competition in which readers could WIN A CARAVAN, details of which could be found on page 24.)

In those days, people were always using the phrase *the New Normal,* though it was unclear how it could ever be normal for the world to be on fire.

The sound of water was a constant presence in our lives at that time—on the bedroom stereo, on Bluetooth speakers, in the car on the motorway. We used it to calm the baby, to send her to sleep. We relied in particular on a playlist called "Ocean Sounds of Martha's Vineyard." These recordings were so omnipresent in our life that we joked about someday taking her to the real Martha's Vineyard to hear her favorite tracks in a live setting. ("Oh, 'Lambert's Cove,' absolute banger," we would say when "Lambert's Cove" came on.) We ourselves were calmed and reassured by the sound of lapping tides, the roaring waves.

But sometimes not. Sometimes, I would be walking up and down the length of our bedroom, holding my daughter to my chest, a soft little animal, the burbling and lapping of the ocean blasting through the stereo speakers, and the sound would become suddenly sinister, and I would imagine seawater rushing up the stairs and into the bedroom, rising up around me as I held her close. The simplest things seemed filled with the urgent purpose of foreshadowing. The catastrophe of the closing act was inscribed into the scenery of our lives.

One Saturday that summer, we were in the park with two friends of ours, sitting on rugs, eating crisps and sandwiches and basking in the lingering heat of the afternoon. They had a baby son not much older than our little girl. The babies were asleep in their buggies, blankets draped downward over the canopies to shade them from the sun. My wife asked our friends if they had any designs on a third child. No, they said. With two, if there was ever some kind of emergency situation and they needed to leave quickly, they could grab one child each and run; with three you were much less mobile. They were sort of joking, but also sort of not.

This idea was one that we returned to now and then through that summer, my wife and I. Should we continue to hold on to certain baby accoutrements, in case we might want to "go again"? There would be a pause, and one of us would say it: with two, we could take one each and run.

There was, in those days, a meme that was ubiquitous on social media, a two-panel cartoon of a dog sitting at a table, the room around him engulfed in flames. In the first panel, the dog is smiling, a coffee cup on the table in front of him. In the second, he is smiling even more resolutely, though the flames are growing nearer, and a speech bubble has him saying "This is fine." I thought about that meme a lot in those days. It came up in a lot of conversations. We would see the wildfires on the edges of Europe, the furious red of the map, the grass turning brownish yellow in the park, and we would say that this was fine, and we would try to mean it, though we knew that we did not.

I mentioned the meme to my therapist one day. She was

not familiar with it—a gap in her knowledge that I found, in the end, quietly reassuring—but she knew what I was getting at in bringing it up.

"Do you ever expose yourself to other kinds of views?" she asked.

"What kind of views?" I said. "That everything might not be fucked?"

"Well, yes," she said. "There is a psychologist, quite famous, who has written a book about how everything has been getting better for humanity over time, and that this is the best moment in history to be alive."

"Are you talking about Steven Pinker?"

"Maybe," she said. "I am terrible with names."

"Voluminous mane of silver ringlets? Looks like Brian May out of Queen? Goes on about the Enlightenment all the time?"

"Yes," she said, "I think that's him."

"I don't find him particularly convincing," I said, more dismissively than I intended to.

She shrugged in a particularly French-seeming way, raising her eyebrows, dipping her head to one side. She was clearly not willing to sacrifice the last fifteen minutes of our session to a defense of Steven Pinker.

There was a silence then, during which I gazed out the window and listened to the bell of an approaching tram, and fell to thinking about Pinker's hair. I couldn't decide whether he had great hair or terrible hair. Like the world itself, I reflected, it depended on the attitude you took toward it. For a prominent champion of Enlightenment values—progress, reason, science, all that—the hair was certainly thematically consistent, in that it looked like one of those powdered wigs men went around wearing in the eighteenth century. I decided that it was resolutely the hair of an optimist, but that in spite of this—or perhaps because of it—it was in fact very bad hair. It struck me then, as I continued to stare out the window, how odd it was that so many of the great pessimist thinkers had, by contrast, terrific hair. I thought of Samuel Beckett, with his incomparable steely crest and his pitiless vision of a meaningless existence, and of E. M. Cioran, who for all his eloquent condemnation of existence as an irrecoverable catastrophe, sported what was surely the most lusciously debonair coiffure in the entire history of philosophy. And then there was Kafka himself, with his great partitioned dome of jet-black hair, established above his high forehead like an auxiliary brain. It was interesting, I thought, how these men managed to maintain both punctiliously styled hair and unremittingly bleak views of human existence.

"What's coming up for you?" asked my therapist.

"Nothing much," I said.

This was a thing that happened often in therapy. There would be a long silence and I would find myself having the most inane and frivolous thoughts—thoughts which, even when pressed to speak about what was on my mind, I found myself reluctant to put into words, for fear it might seem I was taking the whole process insufficiently seriously.

I talked, as I often did in therapy, of my sense of a not-yet-manifested crisis, of the apprehension I nurtured that everything was, in some hazily delineated but nonetheless absolute sense, destined to go to shit. My therapist wanted to know whether this sense of impending crisis was something I was thinking or feeling.

"I find it hard to say," I said. "The distinction between thinking and feeling isn't as clear for me as it is for you."

"But they are two very different things," she insisted. "I notice you make gestures toward your head when you speak of this, which suggests to me that it is something you are thinking."

"Maybe," I said. "But can't you feel things in your head?"

She gave me one of her looks of humorous admonishment. For reasons that were essentially mysterious to me, these facial expressions were among her most effective tools in keeping me from excessive abstraction, from shallow intellectual gamesmanship.

What I felt, I said, could be described as joy, at least a fair amount of the time. I had a family now, two children who with every passing day deepened and strengthened my involvement in the world, my sense of life as a good and worthwhile and even beautiful thing. A couple of weeks ago, I had been at home with flu. I had been in bed all afternoon, I told her, drifting in and out of sleep, when I heard my son sneaking into the room. I remained with my face to the wall, pretending to be fast asleep in the hope that this would discourage him from harassing me with demands that I read him a story or otherwise entertain him. I felt him climbing slowly onto the bed, felt the weight of his small body on the mattress, sure that he was preparing to shock me into full wakefulness by jumping on the bed and shouting. But that wasn't what happened. What happened was that he leaned in gently toward me and kissed me on the back of the head, before climbing back down off the bed and creeping back out of the room and closing the door behind him.

It was, I said to my therapist, the sweetest and most tender thing I'd ever known him to do. I was so disarmed by it I felt that my heart might break with joy. What really got me, I said, was the realization that this was a thing I myself did: I'd come into his room at night to check he hadn't kicked the covers off him in his sleep, hadn't wet the bed, and I would do exactly what he had just done to me—I would kiss him on the top of the head, the back of the neck, and retreat quietly from the room. There was something so beautiful, I said, about the thought that he must have been conscious enough, even in his sleep, to internalize this gesture of protective love, and to return it to me in this way.

"Yes," she said. "Because it is yourself you are seeing. And what you see is that you are doing well as a parent. You are seeing the love you are giving returned to you."

This was exactly right, I said. And yet these feelings of joy and tenderness were always shadowed by the intimation of an unthinkable future, by the anxiety of what might be lying in wait for him and his little sister. I could try to distinguish between feeling and thinking, I said. I could say that the feeling was joy, and the thinking was the shadow that fell across it, but this seemed to me in the end like an artificial separation.

These apocalyptic anxieties of mine—the incessant reading of signs and portents, the perverse fantasies of disaster and collapse—were enfolded in a complex fabric of guilt and self-contempt. Because wasn't the impulse to catastrophize, to imagine the collapse of one's world, only the pursuit of a mind shaped by leisure and economic comfort? What did I really mean by the end of the world, after all, if not the loss of my own position within it? What was it that made me anxious, if not the precariousness of the privilege I had been born to, had passed on with doubtful hands to my own children? In the end, I understood that my fear of the collapse of civilization was really a fear of having to live, or having to die, like those unseen and mostly unconsidered people who sustained what we thought of as civilization. The people who grew the coffee beans for the flat white I bought as I strolled to my office. The factory workers in some gigantic Chinese city, whose name I would never need to know, who made the smartphone on which I listened to leftist political podcasts as I walked, drinking the flat white. The countless homeless people I passed as I walked, for whom civilization had already collapsed, and had perhaps always been so. The end of the world, I knew, was not some remote dystopian fantasy. It was all around. You just had to look.

There are times when I feel as though I am keeping a secret from my son. The reason I feel this way, I suppose, is that keeping a secret is exactly what I am doing. Just as I want him to continue believing in Santa Claus for as long as possible, I want to defer the knowledge that he has been born into a dying world. I want to ward it off like a malediction. I will tell him that he needs to curb his growing addiction to cleaning his ears out at night with cotton buds, because they can't be reused and the plastic is "bad for nature." I will tell him that, for the same reason, we need to be careful about buying too many toys. But I won't tell him that the world is getting hotter all the time, or that the fish are all dying out, or that there may be no real point in planting that last Truffula seed, that the Barbaloots will not be coming back.

I won't tell him such things, because to do so would be in some sense an act of betrayal, even abuse. It isn't true to say that children don't live in the real world. They live in a world that is much realer than ours. The world, to them, is alive. They live into the world, breathing their own presence into everything they see and touch. I have to remind myself not to demystify the world for my son. When he asks me what the moon is doing up there, I have to remember not to tell him that it's a vast dead rock that was likely created by the impact of a massive astral body in the early period of Earth's formation, that it gives no light of its own but rather reflects the light of the sun. It takes an effort of will, an impulse of self-remembering, not to look up the moon's Wikipedia page on my phone, not to give him some desiccated abstract of what I find there. I have to remind myself that my job is not to relay facts, but to keep the mystery alive, to come up with some story or other about the moon, what it is and how it came to be hung there. My job is to maintain for him the realness of the world as he inhabits it. The time for facts will come soon enough, I know.

And the truest fact of all is the damage we have done, are

still doing, to the world. It feels mythological, this business of hidden knowledge, fatal truth. It feels like a manifestation of the doctrine of original sin, the passing down of an immemorial punishment. It is true that the gods are dead, because of course we killed them. But their ghosts are still with us, and the anger of those ghosts is righteous and palpable and poetic.

He asks so many questions, and it is hard to answer them in a way that doesn't reveal too much about the world. This is the anxiety my wife and I cultivate above all others: that we may have already revealed too much, that in our reluctance to talk down to him, we have foreclosed his innocence. On our street is a prison that serves as the national center for male sex offenders. There are, by a presumably large margin, more rapists and child abusers within a two-minute walk of our house than in any other location in the entire country. This fact doesn't exert any kind of urgent psychological pressure: it's not like we're lying in bed at night worrying about the prospect of a pedophile escaping from the prison and breaking into the house. But there is a church beside the prison, and a cemetery that is also a memorial for the republican rebels who were executed after the Easter Rising of 1916, and this is an excellent place for a five-year-old boy to run around and climb trees and race up and down paths on his scooter, and so we take him there often. The walls of the prison loom high over the garden and the trees, and sometimes we kick a ball against them or gather chestnuts beneath them, and there are times, though they are seldom enough, when I remember that the men behind that wall are dangerous men, and that they are there because they did bad things to women and children, up to and including killing them. And in those moments the prison walls, the surveillance tower rising above them, seem like an overbearing metaphor from a world whose existence I am charged with concealing from my son.

One evening my wife and I were talking about our anxious sense of disenchanting his childhood, and she said that she regretted our telling him that the place behind those walls was a prison, and that that was where they put people who did bad things, who hurt other people. (That, for obvious reasons, was about as specific as we got.) We should have made up some story, some other reason for those high walls, the tower. We should have lied to him. I did not disagree. After all, we obsessively guarded his belief in Santa Claus—fretting about his interactions around Christmastime with older children—as though it were a talisman of innocence itself.

In general, we pursued a policy of outright denialism when it came to the broader terror and violence of the world. We brazenly lied to our son, for instance, about the existence of war. War, in the more or less self-contained reality of our house, did not exist—not, at least, until further notice. Wars, we explained, were a thing that was regrettably a feature of "the olden days" but that didn't really happen anymore, people having eventually figured out that they were on balance a terrible idea. This had to be made explicit one rainy Saturday afternoon, because I had taken him to the cinema to see the film *Christopher Robin,* in which Ewan McGregor plays the fortysomething Christopher, charmingly cajoled out of the career-focused torpor of adult life by the sudden rekindling, after a decades-long estrangement, of his friendship with Winnie the Pooh. The pre-credit sequence of this otherwise perfectly harmless film contained a backstory-establishing scene in which the uniformed and knapsacked young Christopher boards a train, presumably in order to fight the Nazis in Europe. Our son's subsequent questions about this aspect of Christopher Robin's life were easily enough settled by appealing to the fact that the film was set in "the olden days," when war was still unfortunately a going concern.

I am confident that we are doing the right thing in pursuing this denialism, but I am aware that it is a policy with a strictly limited shelf life. We know that we are protecting him from things it would do him no good to know, and that this is part of the work of love we call parenting. But there are times when it seems that we are protecting him, and protecting ourselves, from a much deeper and more troubling truth: that the world is no place for a child, no place to have taken an innocent person against their will.

The truth is I don't think enough about the deeper implication of this, of how easy it is to simply turn off the radio when the discussion turns to a bombing campaign in Syria, or to the topic of child abuse in our own country. I rarely think of the sheer randomness of our good fortune, that I can protect him from the horror of the world by reaching for the off button.

That act of switching off is, I realize, not without a certain political friction. Because if I want to teach my children anything, it is precisely not to switch off. What I want to teach them is to listen, to be aware, to consider their relative position in the world, and to be conscious of the ways in which others are less fortunate than they are, and crucially how it might be otherwise. (My own failure to do these things is, of course, an ongoing cause of disappointment and self-censure.)

Intimations of the world's actuality cannot always be fended off. Walking home along the quays one evening that summer, my son and I passed a couple lying unconscious between two frontal columns of the Four Courts, the man sprawled athwart the body of the woman. It was like walking by an open door on a dark street, and glimpsing through it the cold glow of hell itself—another world entirely, a world inside the world. By sheer luck, he was distracted by a couple of pigeons and never turned his little head toward this vision of ruined tenderness.

My son has a dinosaur's tooth mounted in a frame over his bed. We bought it in Edinburgh, in a little shop in the Old Town that sells fossils. The tooth once belonged to a Spinosaur that lived out its days in what is nowadays Morocco. It cost me twenty-five pounds, which seemed surprisingly affordable for something that had been around since the late Cretaceous period. Beneath the tooth, on the white cardboard to which it's affixed, are these words in my own slightly cramped hand: *Dinosaur Tooth, Spinosaurus Aegypticus, 69 Million Years Old.*

Sometimes when I'm kissing my son good night, or reading him a story, I look up at it and am struck by the sheer oddity of its presence in his bedroom, alongside the various pictures around it—a robot, some rabbits, a raccoon playing the fiddle—this curved incisor from a sixty-foot-long carnivore that went extinct ninety-six million years before our own species appeared on the Earth. It's the cheap IKEA frame that does it, I think: this remnant of a former world contained in so representative an artifact of our own. I look at it, and time telescopes forward in my mind, and I imagine the whole thing—ancient tooth, slightly less ancient frame—constituting a single compound relic, perhaps mounted somewhere in yet another frame, in a place and time completely inconceivable from the vantage of our own. *Wood and Plastic Frame, Northern European Origin, 50 million years old; Dinosaur Tooth, 119 Million Years Old.*

And then what comes to mind is the sentence that has haunted me since I heard Caroline come out with it that afternoon as we all sat around the lodge in the highlands: "I wonder whether we humans will make beautiful fossils."

I think about this sentence a great deal, because the world often gives me cause to remember the damage we're doing to it, and to ourselves, with our voracious immolation of fossilized organic matter. For example: we flew to Edinburgh and back with the dinosaur tooth, burning more fossilized organic matter as we went, pumping more carbon into the warming air. The frame was bought for about three euros by a company that sells a hundred million limited-life-span household items every year, single-handedly consuming 1 percent of the world's entire commercial wood supply.

There is no way of contemplating the catastrophe of our way of life from the outside. There is no outside. Here, too, I myself am the contaminant. I myself am the apocalypse of which I speak.

For many years, I considered myself a pessimist. This is not to say that my own experience of life was a miserable one. I was, broadly speaking, a happy and fortunate person for whom the world had laid on a great many privileges and benefits. But to the extent that I could claim to have a basic philosophical position, it was that life, for most people in most places, was characterized by terrible suffering, for no good reason, and that it was unlikely to get any better over time, and that it was therefore on balance probably more trouble than it was worth. Throughout my twenties and into my thirties, the writers who seemed to me to possess the truest vision of the world, who seemed to speak to me out of the deepest wisdom and authority, were those who most firmly denounced the possibility of hope, who rejected most thoroughly the idea that life might be on aggregate a good thing.

In the black gleam of Schopenhauer's prose, I saw a particular reflection of the world's true darkness. Certain passages of his struck me, in those days, with the fierce clarity of a divine decree. Lines like these, from "On the Suffering of the World": "In early youth we sit before the impending course of our life like children at the theatre before the curtain is raised, who sit in happy and excited expectation of the things that are to come. It is a blessing that we do not know what will actually come. For to the man who knows, the children may at times appear to be like innocent delinquents who are condemned not to death, but to life, and have not yet grasped the purport of their sentence."

Who could argue with such a bracing bleakness, such a brave and rigorous rejection of the world? I for one felt no inclination to do so. Every glance at a newspaper, at the scrolling abyss of my Twitter timeline, was a reaffirmation that everything was both as awful as possible and somehow getting steadily worse. Pessimism seemed the only reasonable position to take in relation to it all—to the relentless degradation of the natural world, the wars and the disasters and the random acts of perverse violence and insanity.

A question I have frequently asked myself is whether the appeal of the apocalypse, in all its vastness and finality, is that it can comfortably absorb the personal fear of death. And not just death, either, but every other ancillary fear, too—of change, of instability, of the unknown, and of the precariousness of life itself, all positions held within it.

Given the world, given the situation, the question that remains is whether having children is a statement of hope, an insistence on the beauty and meaningfulness and basic worth of being here, or an act of human sacrifice. Or is it perhaps some convoluted entanglement of both, a sacrifice of the child—by means of incurring its birth—to the ideal of hope? You want to believe that it is you who have done your children a favor by "giving" them life, but the reverse is at least as true, and probably more so.

You want to believe that you are doing a good thing, and that it might mean something to be doing it. And it can't be ruled out, even at this late stage, that you are not wrong to believe these things.

Because the truth is that, for me, the experience of parenthood has meant a radically increased stake in the future. It's not simply that I care about the world in a way I somehow didn't before I had children, but rather that the future has become a realer and more intimate presence in my life, something in relation to which I no longer feel inclined to take abstract positions. I no longer feel the definitive force of pessimism as a philosophy. Statements of hopelessness, no matter how elegantly formulated, no longer sound quite the same tone of authority and wisdom. Which is not to say that I have become an optimist, or anything even close, but simply that life no longer seems to afford me the luxury of submitting to the comfort of despair.

"Optimism and pessimism," wrote Hans Magnus Enzensberger, "are so much sticking plaster for fortune-tellers and the writers of leading articles. The pictures of the future that humanity draws for itself, both positive and negative utopias, have never been unambiguous."

At time of writing, my daughter is almost nine months old. It would be true to say that when I think about her future—by which I mean *the* future—I feel a swelling tide of anxiety within me, that I observe a looping psychic montage of melting ice caps, extreme weather events, fires, droughts, floods, resource wars. I hold her, and look down at the soft indented area at the crown of her head, the fontanel, and see that it is pulsing, and feel suddenly overwhelmed by her fragility, her openness to the world and all its potential harms.

But it would also be true, and maybe even truer, to say that her existence has deepened my investment in the world, my sense of the ever-present possibility of joy, of the future as a fertile realm of possibility, of life.

The thing that goes straight to my heart is how thrilled she clearly is to be alive, this child. She is a tiny engine of joy: the word that comes most readily to mind when I think of her, this beloved and entirely mysterious person, is *refulgence.*

There are songs we sing to her, goofy little ad hoc compositions that have become part of the evolving cultural canon of our family. She knows these songs are about her, because their lyrics are mostly just variations on the theme of her name, and when we sing them she gets so happy that we sometimes worry she might be in some kind of physical danger, that her system might be in some obscure but very real way risking overload. When her brother sings to her, she literally vibrates with joy, with an excess of vital energies, as though in the grip of some sacred ecstasy, ancient and impenetrable. Her chubby little fists clutch the air rhythmically, grasping more of the world, more of what she's feeling, more of us. Her mother, her father, her brother.

There are times when I forget that I'm supposed to be thinking about the end of days, that I'm supposed to be channeling the apocalyptic energies of our time, metabolizing the unease, the fleeting visions of disintegration and dissolution.

There are times when I live only in the present, and it is a good place for the time being.

One evening, toward the end of the weird dry summer of 2018, we were driving home after a visit to my wife's parents. There was very little traffic, and the children were quiet in the back of the car. Our son was playing with a Ninja Turtle, stretching its rubbery arms as far as they would go, sustaining as he played a happy stream of self-contained chatter, half-lucid trash talk, threats, and counter-threats. His sister was asleep beside him. At some point, I became aware that he had stopped talking, that he was being unusually quiet. I glanced in the rearview mirror and saw that he was gazing out the window.

"Look at the sky," he said.

The air was illuminated by the setting sun, by a lurid spillage of purples and pinks and oranges, spreading and deepening. It was one of those spectacles only nature could have successfully pulled off. If anyone else had tried it, it would have looked garish and tasteless. By rights it should have been an aesthetic catastrophe, but somehow it was working.

"It's so beautiful," he said.

His mother and I both agreed that it was lovely.

For a while then we said nothing, and he continued to stare out the window at the blaze of color.

"And it's very *interesting,*" he said.

I waited for him to elaborate, but he seemed content to leave it at that. I had never heard him use the word *interesting* in quite this way before, and I'd never known him to take much note of the sky. This was something new. He was right, I thought: it was very interesting. And I was very glad that he thought so.

It sounds grandly sentimental, and even naive, to say that the world is changed by the birth of a child. But it is also not untrue, or not at least entirely so. It could be said, of course, that it is changed for the worse. It could be said that the pitter-patter of tiny carbon footprints can at this late stage only exacerbate our predicament, and I would not presume to argue otherwise. But are we then expected to hasten the end, to succumb at last to the logic of oblivion, by renouncing the reproductive imperative?

In *The Human Condition,* Hannah Arendt argues that in human affairs it is reasonable to expect the unexpected. As a species, our very existence is such a long shot as to be a kind of miracle, and the fact that organic life exists to begin with is, she points out, an outrageous repudiation of the odds. Everything new in the world enters it miraculously, against near infinite improbability. "And this again is possible," she writes, "because man is unique, so that with each birth something uniquely new comes into the world. With respect to this somebody who is unique it can be truly said that nobody was there before."

This, I think, is the miracle whose dimensions I struggle to take the measure of when I look at my baby daughter, her chubby fists grasping the air, her eyebrows shooting upward in a gesture of inexplicable wit: that nobody was there before, and that somebody is there now, and that it is her.

There is a question I continually put to her, when I am in the mood to question an infant. “Where,” I ask, “did you even *come* from?”

Later, toward the end of the book, Arendt returns to this hopeful business of miracles, of new people, new possibilities. It might be true, she allows, that all human affairs are defined in the end by their movement toward death, that history is a procession in the direction of doom. Death is the only certainty, after all, the only reliable law. And this law, she writes, “would inevitably carry everything human to ruin and destruction if it were not for the faculty of interrupting it and beginning something new, a faculty which is inherent in action like an ever-present reminder that men, though they must die, are not born in order to die but in order to begin.” And it’s the ever-present capacity for new beginnings, she writes, of new births and new people, which is the “miracle that saves the world” from its natural ruin.

It does give me hope to read this, but it is in the end a compromised sort of hope, heavy with guilt and recrimination. It’s the same sort of hope I feel when I lie on my son’s bed, reading the final lines of *The Lorax.* Arendt’s beautiful notion of beginnings in the midst of endings is not unlike the handing down of the last of the Truffula seeds. It allows us some hope, but it puts a lot of pressure on the objects of that hope. Another way to put this might be to say that the reason to be afraid for the future—the children who will have to live in the world we have made, and are still making—is the best reason to have hope for it.

Lately, though, I find that I am no longer submitting to my

own ethical interrogations on the topic of reproduction. I find that I am no longer retrospectively agonizing over whether it was morally wrong to have had children in the first place. The question itself, or the act of asking it, has come to seem essentially absurd. We put on a song she likes, and my daughter's eyes brighten, and she raises her soft little hands in the air and starts to bounce up and down in approximate rhythm with the beat, and suddenly all such philosophical considerations are exposed as frivolous, almost embarrassingly beside the point. Because the point, for the time being, is obvious. The point, more or less, is dancing.

It strikes me now, having extracted it from life and set it down here in the context of this book, that this image—my baby daughter doing her little syncopated bounce, tiny hands aloft—echoes discordantly with the images of those two other dancers: the old man jumping up and down with the crucifix outside the gas station in Los Angeles, the kid head-banging and shredding air guitar on the street in South Dakota. It seems to me that I was employing a form of pathetic fallacy with these guys, projecting my own psychological states onto fleeting visions of complete strangers, people of whose lives or situations or motivations I knew exactly nothing. I was aestheticizing what may have been suffering or madness or ecstasy, and thereby reducing it to a neat emblem of my own anxiety. I was the one channeling the death-seeking energies of the culture, dancing the ghost dance of the mind. They were simply dancing. They were dancing, more than likely, for the same reason my daughter dances. Because they felt like it, because they happened to be alive, and really what else is a person who finds themselves in that situation supposed to do?

I will say that at the time I began writing this book, I was preoccupied by visions of catastrophe, instinctive fantasies of retreat. All those hours watching prepper videos on YouTube, thinking about bunkers and food storage and water filtration and so on: my approach to this was intellectual and in all senses of the word critical, but on some level I always understood my irony to be also a pose, also a kind of defensive crouch. One man reads the signs and portents, picks up the unmistakable scent of blood in the air, and he builds a bunker, stocks up on flavored protein sludge, makes a YouTube video about surviving civilizational collapse. Another receives these same signals and meditates on the meaning of the bunker, the protein sludge, the man in the video, etc., etc., etc. Both are looking for ways to negotiate their terror.

I wonder now whether it is because of or despite the strange series of pilgrimages I have been on that I have come at last to this place of accommodation, tentative though it may be. How could it be that after more than a year in search of vistas of devastation, intimations of the end, I no longer feel such despair about the future? I am tempted to say that I cured myself of my apocalyptic anxiety by means of a kind of exposure therapy. Though I suspect there might be some measure of truth to this, it feels diluted to a homeopathic degree. No, the real truth, as always, is simpler, and as always it is more mysterious.

Somewhere along the way, in any case, it became apparent to me that a state of perpetual anxiety was no way to live. It became apparent that my obsessing over the end of the world constituted a kind of retreat, and that that retreat was a kind of dying. "The taste for worse case scenarios," wrote Susan Sontag, "reflects the need to master fear of what is felt to be uncontrollable. It also expresses an imaginative complicity with disaster." I knew the first claim to be accurate, and suspected the second was not far from the truth either.

The future is a source of fear not because we know what will happen, and that it will be terrible, but because we know so little, and have so little control. The apocalyptic sensibility, the apocalyptic style, is seductive because it offers a way out of this situation: it vaults us over the epistemological chasm of the future, clear into a final destination, the end of all things. Out of the murk of time emerges the clear shape of a vision, a revelation, and you can see at last where the whole mess is headed. All of it—history, politics, struggle, life—is near to an end, and the relief is palpable.

I have known my own moments of cosmic nihilism. I know how it feels to consider the prospect of total destruction, the annihilation of all human meaning, and to take comfort in saying *so be it, let it happen.* I have felt that strange peace, watching a pair of kestrels spiraling upward through the shell of a cooling tower, black against the inhuman blue of a Chernobyl sky. I have felt it in the Scottish Highlands, attuning my ears to the sound of a world without human voices, and even in a windowless room in Pasadena, listening to knowledgeable men talk about our destiny as a multi-planetary species, thinking to hell with the future and everything else. I have felt it at this very desk, watching streaming footage of a floating trash continent in the Pacific, the Great Barrier Reef in a state of noxious decay.

Science, for what it's worth, is unambiguous on this point: all systems inexorably tend toward total entropy. Ice caps, political orders, ecologies, civilizations, human bodies, the universe itself. In the long run, everything is nothing.

But in the meantime, everything is not nothing, not even close. In the meantime, we have no idea what might come to pass. Make of that what you will, is my point.

And so lately, I have lost my taste for cosmic nihilism, cosmic despair. Lately I have been glad to be alive in this time, if only because there is no other time in which it's possible to be alive. And I would think it a real shame, in the end, if there were nobody around to experience the world, because although it is a lot of other things, I would agree with my son that it is also an undeniably interesting place. You kind of have to hand it to the world, in that sense.

I've been working recently in the reading room of the National Library on Kildare Street. The building is adjacent to Leinster House, where Ireland's houses of parliament sit. The other day, I heard the sound of chanting drifting up from the street, high and insistent, but couldn't quite make out what the chanters were saying, and so I closed my laptop and walked down the marble staircase and out into the street. I didn't stop to grab my coat from my locker, and I didn't regret it either, because although it was only February it was wasn't strictly speaking jacket weather. The day was bright, and it was warmer than it had any business being this time of year. Intellectually, I understood this to be an omen of apocalypse, but I felt it as the unexpected blessing of an early spring, a day suddenly warm with life and possibility. Maybe it was the end of the world, or maybe it was just a nice day, or maybe it was both.

Across the street was a crowd of children, perhaps a couple hundred. They were primary school kids, the youngest of them not much older than my own son. They were chanting for climate action, holding up placards that they had made themselves, presumably in class. One girl held a sheet of paper bearing a crayon drawing of the planet with a sad little cartoon face on it. Another depicted a gigantic blazing sun in yellow and red crayon, beneath which was a somewhat weirdly proportioned man, sweating lavishly, a speech bubble saying IT'S TOO HOT! One kid held up a drawing of a dinosaur skeleton, a message in even-handed capitals that read: THE DINOSAURS THOUGHT THEY HAD TIME TOO. Another had a sign on which he'd scrawled the impressively enigmatic words EARTH IS FALLING . . .

I walked up the street a little distance and stood watching them awhile. They were all wrapped up in coats despite the relative warmth of the day, some of them in woolen hats and scarves. I thought of all their mums and dads, wrapping them up against the elements, buttoning their coats up to the top, telling them not to forget their drawings, their placards. I looked at their faces and was touched by the innocence and exuberance I saw there, the total absence of self-consciousness or cynicism. They seemed to me to be fully inhabiting themselves, to be fully and unequivocally alive.

I muttered quietly to myself the word *unless.* It was unclear to me why I had said this word, and I was surprised to hear myself laughing, feeling as I did so a strange and volatile mixture of sadness and delight.

*Unless someone like you cares a whole awful lot, nothing is going to get better. It's not.*

At home that evening, I was watching my wife breast-feed our daughter, propped up on our bed against a pile of pillows. The baby was in a sleeping bag, which for some reason had images of little colored cartoon bugs all over it, and I was struck by the incongruity of an item of baby apparel featuring creepie crawlies, harmless and friendly-looking ones though they were.

"Did you hear about how insects are going extinct?" I said to my wife.

"No," she said, looking up from the top of the baby's head. "I hadn't heard that. Which ones?"

"All of them," I said. "As, like, a category. There was a thing in *The Guardian* about it today."

"Jesus," she said, looking momentarily forlorn.

I found the article on my phone and scanned through it for choice phrases.

"Hurtling down the path to extinction," I read.

And then: "Leading to a catastrophic collapse of nature's ecosystems."

Even as I read these phrases, I wondered what I was up to. It seemed a strangely indecent thing to be doing, standing there reading an incoherent sermon of apocalypse while my wife tried to breast-feed our daughter.

"Weren't the insects supposed to be our fallback protein

source," she said, "once we all stop eating meat because of climate change?"

"Were they?" I said, locking my phone screen and flinging it wristily onto the bed. "That does ring a bell. Insects, though, Jesus. If they go, we all go."

I had no idea whether this was true. Maybe there was some way we could continue to live in a world without insects. It seemed like a long shot, though.

"That's a worry," said my wife, wincing suddenly as the baby bit down on her. This was a thing she'd been doing lately. She seemed to think it was funny. She'd bite down on the nipple she was feeding from, and then look up at my wife to gauge the reaction. She was turning into something of a joker, our little girl.

I hoped she would be a happy person, resilient and resourceful. I hoped she would be innocent for as long as she needed to be. I thought of the protesting kids earlier, all wrapped up for a day that was a lot warmer than it should have been.

I reached out to her, feeling under my hand the gentle blonde fuzz, the impossible softness of her little head. She pulled abruptly away from my wife and turned her face toward me. She presented me with an impish look, an expression of mock seriousness, and blew a raspberry in my direction, impressively loud and sustained. I laughed and blew one back at her, which seemed to be the desired response.

# ACKNOWLEDGMENTS

My gratitude is due to the following people: Amy Smith, Molly Atlas, Karolina Sutton, Yaniv Soha, Anne Meadows, Lamorna Elmer, Dan Kois, Cara Reilly, Matteo Codignola, Roberto Calasso, Benedetta Senin, Francesca Marson, David Wolf, Max Harris, Andrew Dean, Matt Nippert, Khylee Quince, Peter Skilling, Annie Goldson, Jonathan Shainin, Max Porter, Anthony Byrt, Dylan Collins, Michael and Deirdre O'Connell, Ed Caesar, Bush Moukarzel, Norah Campbell, Simon Denny, Lisa Coen, Sarah Davis-Goff, Andres Roberts, Paul Kingsnorth, Susan Cross, Caroline Ross, and Ronan Perceval.

I am also grateful to the Pulitzer Center for generously funding my travel to New Zealand.

# ABOUT THE AUTHOR

Mark O'Connell is the author of *To Be a Machine,* which was awarded the 2019 Rooney Prize for Irish Literature and the 2018 Wellcome Book Prize and short-listed for the Baillie Gifford Prize for Non-Fiction. He is a contributor to *The New York Times Magazine, Slate,* and *The Guardian.* He lives in Dublin with his family.